PARKSTONE LIBRARY
Britannia Road, Parkstone
Poole BH14 8AZ
Tel: 01202 742218
Email: parkstonelibrary@poole.gov.uk
1 2 SEP 201
FOR SALE
WITHDRAWN
FROM STOCK
KT-513-879

boroughofpoole.com

SILENT WITNESSES

SILENT WITNESSES

THE STORY OF FORENSIC SCIENCE

NIGEL McCRERY

arrow books

Published by Arrow Books 2014
2 4 6 8 10 9 7 5 3 1

First published in Great Britain in 2013 by Random House Books
Random House, 20 Vauxhall Bridge Road,
London SW1V 2SA

www.randomhouse.co.uk

Addresses for companies within The Random House Group Limited can be found at:
www.randomhouse.co.uk/offices.htm

The Random House Group Limited Reg. No. 954009

A CIP catalogue record for this book
is available from the British Library

ISBN 9780099569244

The Random House Group Limited supports the Forest Stewardship Council® (FSC®), the leading international forest-certification organisation. Our books carrying the FSC label are printed on FSC®-certified paper. FSC is the only forest-certification scheme supported by the leading environmental organisations, including Greenpeace. Our paper procurement policy can be found at www.randomhouse.co.uk/environment

Typeset in Adobe Garamond Pro by Palimpsest Book Production Limited,
Falkirk, Stirlingshire

Printed and bound by CPI Group (UK) Ltd, Croydon CR0 4YY

Dedicated to Professor Helen Whitwell, my friend and the inspiration for the series *Silent Witness*. The real Sam Ryan.

Contents

Introduction

Murder has a magic all of its own.

William Roughhead, Scottish criminologist (1870–1952)

The morning of 21 November 1983 dawned cold. The wind was bitter, the sky dark and bleak. On her mother's advice, fifteen-year-old Lynda Mann dressed warmly before leaving for school. She wore denim jeans over a pair of tights, a thick pullover, white socks and black tennis shoes. Before leaving the house she also pulled on her new donkey jacket and stuffed a warm scarf into her pocket.

Lynda lived in Narborough, a village about six miles from Leicester city centre. It was what Lynda's mother Kathleen described as 'a real English village'. Divorcée Kathleen had been a city dweller for most of her life, but had settled there with Lynda and her other daughter, Susan, after falling in love with the place. In 1980 she married Eddie Eastwood, a former soldier, and they became a happy family of four.

Lynda herself was an attractive, dark-haired girl with pale

skin. She was outgoing, bubbly and enthusiastic. She was doing well at school, was studying several languages and was determined to travel widely as soon as she was able. She seemed to love life. As so often seems to be the case in situations like this, she didn't have an enemy in the world.

After school that day, Lynda returned home for a quick meal with her stepfather before going back out into the village. She went to visit a friend called Karen Blackwell for a short while, before moving on to the home of another friend to collect a record she had lent her. This girl, Caroline, lived in Enderby, a fifteen-minute walk from Karen Blackwell's, close to a secluded footpath known locally as the Black Pad. It was as Lynda made her way back from here that she noticed a figure standing by a lamppost, not far from the gate to the Carlton Hayes psychiatric hospital.

At 1.30 A.M., Lynda had still not returned home. Growing increasingly concerned, her stepfather drove around the village in search of her. He visited various local hangouts including the Black Pad. When this proved fruitless he went to Braunstone Police Station and reported Lynda missing. The police took down her details but weren't overly concerned since she hadn't been missing for very long. Eddie Eastwood then went back home to wait. What he didn't know was that when he was searching the Black Pad, he had been only feet away from making a horrible discovery.

The following morning a hospital porter on his way to work decided to take a short cut across the Black Pad. As he did so he noticed what he thought at first was a partly clothed mannequin lying on the grass near a clump of trees. The body was as white as marble and rigid. As he approached he realised that it wasn't a dummy at all; it was a young girl. He had discovered the body of Lynda Mann.

The police were called and Detective Chief Superintendent David Baker attended. At 8.30 A.M. on 22 November 1983, the murder inquiry had officially begun.

The case would go on to become a landmark in the history of forensic science. By a strange coincidence, the technology that was to prove decisive in solving it was developed only a few miles from Narborough, at the University of Leicester, roughly a year after Lynda's tragic death.

Dr Alec Jeffreys (now Sir Alec) was a graduate of Merton College, Oxford, where he read biochemistry. He remained in Oxford to study for his PhD and, after receiving it, worked for a short spell as a research fellow at the University of Amsterdam, before moving to the University of Leicester in 1977.

It was on 10 September 1984 that Jeffreys made a revolutionary discovery. While examining an X-ray film image of a DNA experiment, he happened to notice that the DNA of different members of his technician's family showed both significant similarities and significant differences. Jeffreys quickly realised the importance of this: that individuals could be identified by the unique variations in their genetic code. Every person has their own genetic 'fingerprint'. This meant that any genetic material – such as hair, skin cells or bodily fluids – could now theoretically be matched with the person from whom it came.

When Lynda's body was discovered, a pathologist was called to the scene. During the course of their examination they noted 'matted seminal stains on the vulva hair' – something that would later come to be highly significant. After she had been identified by her stepfather, a postmortem was carried out. It was established that intercourse had been attempted and that premature ejaculation had occurred. Penetration had also taken place after this and prior to death. Semen was recovered from a deep vaginal swab. The official cause of death was recorded as asphyxia due to strangulation.

The semen was subjected to a phosphoglucomutase (PGM) grouping test. It was also antigen-tested and found to have come from a blood group A secretor; someone from blood group A who secretes antigens from their blood into other bodily fluids such as semen or saliva. The science here is complicated; it is enough for our purposes to know that this meant that the killer was a Group A secretor PGM1+. This was the first breakthrough, as this description would apply to only one male in ten in the UK. On its own, this information couldn't absolutely identify the killer, but it was useful as it allowed the police to eliminate suspects – Eddie Eastwood's innocence was confirmed in this way, for example (he was never really under suspicion but in such cases the immediate family always needs to be checked out). However, they seemed no closer to catching the culprit. Leads came and went, suspects were interviewed and allowed to go free. The investigation went on.

Lynda's body was finally released and she was buried at All Saints Church on 2 February 1984. By April that year, the number of active officers on the case had fallen from 150 to eight. The incident room was closed and in the summer the inquiry was run down altogether. During the inquiry, 150 blood tests had been carried out, but they had all come to nothing.

As time moved on, although the memory of Lynda Mann didn't disappear from the consciousness of the village, it did dim a little. The fact that nobody had been held to account kept awareness of the murder alive, but equally the fact that there had been no further incidents helped the tragic occurrences of November 1983 seem more distant. In July 1986, all that was to change in the most tragic of circumstances.

Robin and Barbara Ashworth lived with their two children, Dawn and Andrew, in the village of Enderby, near Narborough. They were a close, loving family. Dawn was fifteen years old and had bright, expressive hazel eyes. She was not particularly academic but possessed a strong artistic streak. To supplement her pocket money, Dawn had a part-time job working in a newsagent.

At 3.30 P.M. on 31 July 1986, Dawn came home after work. She changed quickly and was about to head back out to see her friends when her mother reminded her that she had to be home by 7 P.M. as they were going to the birthday party of a family friend. As a result of this, Dawn decided to go to buy some sweets as a gift. When she left the house she was wearing a white polo-neck pullover covered by a loose-fitting multicoloured blouse with a white flaring skirt and white canvas pumps. She was also carrying a blue denim jacket.

Dawn bought the sweets and her friends last saw her at approximately 4 P.M. heading towards Ten Pound Lane, a country path that was a short cut between Enderby and Narborough. On her way she called on several friends, only to discover from their families that they were already out. If only this had not been the case, an awful tragedy might have been averted. I have read of – and been involved in – many cases where chance has played a significant role in the way that events unfolded. It is a powerful governing force in all our lives – and deaths.

Dawn began to make her way home along Ten Pound Lane.

When she wasn't home by 7 P.M. to attend the party, her parents began to worry. It wasn't like her to be late; she was normally very reliable. Her mother discovered that she had left her friend's house at 4.30 P.M. and hadn't been seen since, which increased their concern. They reported her missing to the police but were told to wait a little longer – it wasn't unusual for a teenage girl to go missing for a few hours. Dawn's parents knew that, in her case, it was.

By 9.30 P.M. there was still no sign of her, and her father went out to search for her. He scoured the local streets and the footpaths and, just like Eddie Eastwood three years before, walked past the very spot where his daughter was lying without seeing her.

The following day, Friday 1 August, the police finally took action, and the Narborough area was alive with search teams and dogs.

As is normal in such cases, both Robin and Barbara Ashworth were interviewed at length and their house and garden carefully searched. During this time they were also subjected to anonymous silent phone calls, adding to their anguish. The papers were full of the search and included a personal plea from her father for Dawn to be returned safely.

On 2 August, a police sergeant discovered a denim jacket close to Ten Pound Lane with a lipstick and a packet of cigarettes in the pocket. The area was immediately sealed off and before noon a body was discovered by a clump of blackthorn bushes next to Ten Pound Lane. The body was naked from the waist down, just like Lynda Mann's had been. The police knew at once who they had found, though it fell to her father to make the official identification. With this done, at 6.30 P.M.

the postmortem commenced. The pathologist established the cause of death as asphyxia due to manual strangulation, probably by having an arm hooked around her throat. She had been raped and sodomised, most likely after death. It was also established that Dawn had been a virgin before the attack.

The inquiry followed the normal pattern: interviews, door-to-door enquiries, re-enactments, appeals. As the police sifted through the intelligence they had gathered, they realised that they had a promising lead. At least four witnesses had reported seeing a man on a red motorcycle or wearing a red crash helmet. Sightings of this man and his bike were made at various times and in various places. He was seen under a nearby bridge at noon, and a different witness saw him there again at around quarter to five. A third witness saw the bike on Ten Pound Lane at 5.15 P.M. and a fourth reported seeing the bike being ridden up and down Mill Lane on the evening Dawn's body was discovered, as though the rider were taking a keen interest in the inquiry.

A seventeen-year-old boy, who worked as a hospital porter at the Carlton Hayes Hospital, was seen by a local police officer pushing a motorcycle. He was stopped and, after he admitted having seen Dawn shortly before she disappeared, was brought in for questioning.

On the following Thursday, 7 August, a witness contacted the inquiry team and told them that the same boy, who was his colleague at the Carlton Hayes Hospital, had told him that the police had discovered Dawn's body in a hedge by the M1 bridge, hanging from a tree. While this last detail was not true, the rest of the description was uncannily accurate, considering that the police had not yet released this information. Another witness then came forward and explained that the boy had told him, only hours after it happened, that Dawn's body had been discovered, again before the police had made an official announcement. It was also alleged that he had acted inappropriately with several women in the past, and that he had told one he was the last person to see Dawn Ashworth alive. One of these witnesses had also noticed scratch marks on his hand when they spoke.

As a result of all this information, Detective Sergeant Dawe and Detective Constable Cooke from the inquiry visited the boy at his house in Narborough and arrested him in connection with the murder of Dawn Ashworth. He was driven to Wigstone Police Station where he underwent a series of interviews conducted by various members of the inquiry team. Over many hours he was gradually worn down until at last he admitted to the murder of Dawn Ashworth. Many of his admissions were contradictory and more than a little vague, but when he was eventually presented with a statement admitting that he had carried out the murder, he signed it. He was then removed to Winson Green Prison in Birmingham.

With her killer safely behind bars, four weeks after her murder, Dawn Amanda Ashworth was finally laid to rest in the churchyard of St John's Baptist Church in Enderby.

Now that they were sure they had their man, the police wanted to make a definite link between Dawn's murder and that of Lynda Mann. It was something the press had already been speculating about. However, there were flaws in the case against the boy. He had given blood and it was quickly established that he was not a Group A secretor PGM1+, something the police had placed a great deal of emphasis on when looking for the killer. But a forensic scientist reassured them by telling them that they were only dealing with maybes, and suggesting such things were perhaps not a 'precise' science. The boy's mother had given him a strong alibi for the evening of Dawn's murder, but this was also dismissed on the grounds that she was a far from disinterested party. In retrospect it seems likely that the police were so relieved to have someone locked up for the crime, and so swayed by the circumstantial evidence against him, that they ignored what were actually real problems with the case.

Exactly what happened next is open to debate. In the end it depends who you believe. The boy's father maintains he had heard of the development of genetic fingerprinting and asked his son's solicitor to look into it. The police, on the other hand, maintain it was their idea to try to prove once and for all that they had the right man. As a result, it will never be clear who put forward the idea of using this new technology in the case, but put forward it was. Dr Alec Jeffreys' work came into play. This was to be the decisive development in the cases of both Lynda Mann and Dawn Ashworth.

Before the murders, Jeffreys had already made legal history by proving through genetic fingerprinting that a French teenager was the father of an English divorcée's child. He was well known and highly respected within the scientific community, but not particularly recognised outside of that sphere. That was about to change.

A senior detective from the Leicestershire Constabulary asked Jeffreys to analyse samples of blood from the self-confessed murderer of Dawn Ashworth, 'just to be sure'. He explained to Jeffreys that the police hoped to prove that the boy had also murdered Lynda Mann.

Jeffreys was given a semen sample from the Lynda Mann investigation. It was somewhat degraded but nevertheless he ran it through his usual process and hoped for the best. Luckily, they were able to obtain a proper DNA profile. 'And there,' Jeffreys recalled later, 'we could see the signature of the rapist.' More importantly, 'It was not the person whose blood sample was given to me.' Jeffreys then went on to spend a week analysing samples collected from the Dawn Ashworth murder.

When he finally had the results, he contacted Chief Superintendent David Baker and told him that he had both good news and bad news. Baker wanted the bad news first. Jeffreys told him, 'Not only is your man innocent in the Mann case, he isn't even the man who killed Dawn Ashworth.' After the detective had finished using the Queen's English in a particularly colourful way, he asked Jeffreys for the good news. 'You only have to catch one killer. The same man murdered both girls.' Baker wanted to know if there could have been a mistake. Jeffreys was firm on this point. 'Not if you've given me the correct samples.'

The boy appeared in Leicester Crown Court on 21 November 1986. It was a day on which both legal and forensic history was made. He became the first person ever to be set free on the evidence of a DNA test. To this day, nobody is entirely sure why he confessed to the crime in the first place, or indeed how it was that he seemed to know so many privileged details of it. It seems likely that he simply caved in to pressure under interview, and that the information he had came from rumours he'd heard and repeated; it just happened to be uncomfortably close to the truth. His acquittal was a triumph for Jeffreys and for forensic science, and an enormous relief for the boy and his family. For the Leicestershire Constabulary, however, it was a disaster. They had no choice but to begin their hunt once more.

They began to search for the real culprit with renewed urgency. A reward of £20,000 was offered for information leading to the arrest and conviction of the murderer, and a fifty-man squad was assembled at Wigston Police Station.

Then, at the beginning of 1987, a remarkable and, it has to be acknowledged, brave decision was made by the inquiry's senior investigation team. They decided to take blood from every male member of the local community aged between fourteen and thirty-one who was unalibied, and from all males who had worked in, or had some other connection with, the villages of Narborough, Littlethorpe or Enderby (although this was later amended to any male born between 1 January 1953 and 31 December 1970 who lived, worked or had a recreational reason for being in the area). This included past and present patients and employees of the Carlton Hayes Hospital.

'The blooding', as it became known, took place in two locations, three days a week, between 7 A.M. and 9 P.M. There was also a late session between 9.30 P.M. and 11.30 P.M. once a week. By the end of January there had been a 90 per cent response and over a thousand men had given blood. However, only a quarter of them had been cleared through testing. The process was obviously going to take longer than the two months initially estimated.

January was a bad month for Colin Pitchfork. He was feeling troubled and having difficulty sleeping. His concerns had started when he received a letter from the Leicestershire Constabulary requesting that he go to one of their clinics and voluntarily give blood. It gave him a time and date to attend. When his wife asked why he was so agitated about it, he explained that he was convinced that the police were going to set him up because he had a previous conviction for indecent exposure. He didn't go.

When the second request arrived, Pitchfork started to approach friends and colleagues at Hampshires Bakery where he worked, offering them £200 if they would take the blood test for him. He cited his conviction for flashing and his hatred of the police as reasons. To their credit, most of his colleagues refused. That was until he approached Ian Kelly. Kelly was a twenty-four-year-old oven man at the bakery and had only worked there for six months. He and Pitchfork were not on particularly friendly terms, but they got on well enough.

Pitchfork took a different tack when trying to persuade Kelly. He told him that he had already given blood for a friend who was scared of getting into trouble because of a previous conviction for flashing and robbery. There was, he said, no chance this friend could have any connection with the murders because he wasn't even living in the village when they were committed. Now he, Pitchfork, was in trouble because he had done an innocent friend a favour. If discovered, his act of friendship might even land him in prison. The next occasion Pitchfork was due to give blood was 27 January. Time was running out for him. He continued to put pressure on Kelly until he eventually agreed to give blood on Pitchfork's behalf.

The whole arrangement nearly fell through when Kelly was taken ill on the day he was supposed to attend the appointment. However, Pitchfork managed to talk him out of his sick bed and the two of them made their way to Danemill School on Mill Lane in Enderby, where blood was being taken (oddly, the school was on the street where Dawn Ashworth had lived). While Kelly gave blood, Pitchfork waited outside, standing in the shadows so as not to be noticed. Kelly did all that had been asked of him; he signed the consent form, and gave both blood and saliva. The job was done.

By the end of May there had been an amazing 98 per cent response to the call for samples. However, of the 3,653 men and boys that had been blooded, only 2,000 had been eliminated due to the laboratory's unusually heavy workload. By now the murder squad had been scaled down to twenty-four officers, and they had over a thousand people still to contact. Shortly after this, the squad was cut again, to sixteen officers. It was left to Inspectors Derek Pearce and Mick Thomas to fight the inquiry's corner against those who wanted to run it down completely.

The breakthrough came, as is often the case, from an unguarded moment on the part of someone involved. One lunch break, Ian Kelly went to the Clarendon pub and met some of his colleagues from Hampshires Bakery. One way or another, the conversation turned to Colin Pitchfork and his inappropriate behaviour towards women. During this conversation, Ian Kelly mentioned that he had given blood for Colin Pitchfork once. When he was asked why, he told them about the murder inquiry. Another of the bakers then mentioned that Pitchfork had offered him £200 to take the blood test but that he had refused.

One of the women there was profoundly disturbed by what she had heard. She asked one of the bakers what they should do about Pitchfork. The reply was simple, 'Nothing.' Everyone seemed sure he wasn't guilty of anything. Besides, it would get Ian Kelly into serious trouble, and nobody wanted that. In spite of this, the woman wouldn't let the matter drop. She discovered that the landlord of the Clarendon pub had a policeman for a son, and she decided that she had to pass the information on, though it was several weeks before she finally contacted the young constable.

When they received this information, the first thing the team did was compare Pitchfork's signature on the house-to-house pro forma from the Lynda Mann inquiry with that from his blooding in January of that year. The two didn't match. On the morning of 19 September, Ian Kelly was arrested by Detective Inspector Derek Pearce for conspiracy to pervert the course of justice. He was taken to Wigstone Police Station to be interviewed. He didn't hold back, telling the police everything they needed to know and naming Pitchfork as the man he had given blood for. For the first time in many months, the team started to become excited.

At 5.45 P.M. the same day, detectives visited Colin Pitchfork's house. They identified themselves and were allowed inside. They took Pitchfork into the kitchen alone and informed him that 'From inquiries we've made we believe you're responsible for the murder of Dawn Ashworth on 31 July 1986.' They also told him that they were aware that someone had given blood for him. All Pitchfork said was, 'First give me a few minutes to speak to my wife.' As he was leaving the room, one of the detectives asked, 'Why Dawn Ashworth?' Pitchfork turned and replied, 'She was there, I was there.' Although the police now felt certain that they had their man, they were also aware that they had made a mistake before. It was down to Jeffreys once again to provide the final proof. This time the DNA test came back positive: Pitchfork really was the murderer of both Dawn Ashworth and Lynda Mann.

Pitchfork made a full and detailed confession. He was tried at Leicester Crown Court on 22 January 1988. He was given a double life sentence for the murders, a ten-year sentence for each of the rapes, and three years each for sexual assaults committed in 1979 and 1985, plus a further three years for the conspiracy involving Ian Kelly. When he gave his sentence, the judge, Mr Justice Otton, commented, 'The rapes and murders were of a particularly sadistic kind. And if it wasn't for DNA, you might still be at large today and other women would be in danger.'

DNA testing, the greatest advance in forensic science for over a hundred years, had come of age. It would go on to affect the outcome of criminal cases around the world; its importance in establishing guilt or innocence cannot be overstated. Today, despite concerns and challenges, it is clear that Dr Jeffreys' remarkable discovery is here to stay.

The case of Lynda Mann and Dawn Ashworth serves to show just what a powerful tool genetic fingerprinting is for a forensic scientist – it offers perhaps the most incontrovertible proof of a person's connection to a scene. But there are, of course, many other techniques at an investigator's disposal. Innovations and advances are continually being made in this field. It is this incredible variety of approaches that makes the history of forensic science such a fascinating subject. For each forensic technique, from ballistic analysis to old-fashioned fingerprinting, there are cases that highlight the real practical value of new developments. In this book I look at some of the most important of these cases, and through them demonstrate that a person still has a story to tell long after they are dead.

I

Identity

Always remember that you are absolutely unique. Just like everyone else.

Margaret Mead, US anthropologist (1901–78)

Forensic investigation is concerned primarily with piecing together the disparate clues left at a scene in order to form a coherent picture of events and, crucially, to establish the identities of those involved or – equally importantly – those who were not. However, it wasn't until the nineteenth century that the need for a reliable, systematised method of identifying the people involved in a crime was recognised. Prior to then, the most common ways of doing so were eyewitness accounts and information extracted by torture. Needless to say, both could easily provide a faulty account; as this was recognised, various experts rose to the challenge of improving matters. The pioneering French forensics expert Edmond Locard (1877–1966) once said that 'to write a history of identification is to write the history of criminality', and of course most forensic science is concerned either with establishing identity or with linking an individual to a crime scene. This chapter looks at the first, most basic steps in this direction – the early attempts to define and catalogue a person's physical characteristics. There was a pressing need to formalise methods of identification, as the case of Lesurques and Dubosq in France showed.

On 27 April 1796, the Lyon mail coach failed to arrive in Melun, a small hamlet south of Paris. Concerned, the people of Melun assembled a search party. It did not take them long to discover the coach, and the sight that greeted them was a gruesome one. Both the driver and the postboy had been hacked to death and their bodies badly mutilated. The apparent motive for the crime became clear when it was found that more than five million francs had been stolen from the coach. One of the horses had also been taken.

Since the coach's only passenger was not among the dead and had, in fact, completely vanished, it seemed pretty clear to the authorities that he had been part of the gang that had committed the murderous robbery. He had claimed to be a wine merchant, but in fact must have been acting as the gang's inside man all along. It also came to light that he had been seen prior to boarding the coach carrying a large cavalry sword; given the condition of the bodies it seemed that this might very well have been used as one of the murder weapons. After a short investigation it was established that the gang likely comprised four other members, who had also been heavily armed – a gang of four such men had eaten in the nearby village of Montgeron a few hours before the coach was due to arrive there, and had been acting suspiciously.

The police quickly picked up the gang's scent. The missing horse, which had been taken from the coach, was discovered in Paris the following day, and not long afterwards a stable-keeper reported that four sweating horses had been returned to his stable during the early hours of the morning by a man who gave his name as Couriol. Couriol was eventually traced to a village just north of Paris and arrested. Both he and his premises were searched and over a million francs recovered. The police were convinced that they had their man, and he was taken to Paris to answer further questions and be put before the Palais de Justice. The case then took an unusual turn.

A man by the name of Charles Guenot had been found in the same house as Couriol. Although after questioning him they had decided he was not a suspect, the police had taken some papers from him. As a result, Guenot was forced to go to Paris the following day in order to retrieve them. On his way he bumped into an old friend by the name of Joseph Lesurques, a rich businessman from Douai. Guenot explained what had happened and Lesurques, sympathising with his situation, agreed to go with him. By a strange coincidence the two barmaids from Montgeron who had served the gang their meal on the fateful day were also there, helping with the inquiry. When they saw Guenot and Lesurques together, they pointed at them and denounced them, convinced that they recognised them both as members of the group.

Guenot and Lesurques were immediately arrested on the basis of this evidence. Despite fervently protesting their innocence, they were tried along with Couriol and three other men who were accused of being accomplices. Guenot was acquitted, but all the other men, including the hapless Lesurques, were found guilty and sentenced to death. The conviction of Lesurques seemed especially bizarre considering that no fewer than fifteen witnesses provided him with an alibi, while a further eighty-three spoke highly of his character and respectability. For some reason all this evidence was ignored by the court and the evidence of the two women, who never wavered in their account and their identification of Lesurques as one of the men who had attacked the coach, carried the day.

On hearing himself condemned, Lesurques, who had remained confident and assured throughout the trial, finally lost his self-control. Raising his hands to the heavens he declared: 'The crime which is imputed to me is indeed atrocious and deserves death; but if it is horrible to murder on the high road it is no less so to abuse the law and convict an innocent man. A day will come when my innocence will be recognised, and then may my blood fall upon the jurors who have so lightly convicted me, and on the judges who have influenced their decision.'

Immediately after the trial, in an act of contrition Couriol, who was indeed guilty, made it clear that Lesurques really was completely innocent and had taken no part whatsoever in the crime. The judge who had ordered Lesurques' arrest, a man by the name of Daubanton, was so disturbed by this revelation that he went to see Couriol in prison to speak to him personally. Couriol stuck to his story, explaining that the waitresses were wrong and had mistaken Lesurques for the real culprit, a man by the name of Dubosq who looked similar. The major difference between the two men was that Dubosq, unlike Lesurques, had dark hair. However, at the time of the robbery (and for some time beforehand), Dubosq had worn a blond wig in order to disguise himself.

To his credit, Daubanton had the case reopened and a commission was established to re-examine the evidence against Lesurques.

It was pointed out to them that Lesurques had no possible motive to get involved in highway robbery, as he was already rich. He was also, as we have already noted, very respectable – not the kind of man likely to carry a heavy sword around with him, or to have any idea how to use it if he did. However, in an extraordinary piece of deduction, the commission decided that perhaps Lesurques' relatives had bribed Couriol's relatives in order to persuade him to declare Lesurques innocent. Despite there being no evidence whatsoever to support this ridiculous theory, the Minister of Justice agreed and the sentence of death was upheld. Given the insanity of this decision, I for one have often wondered whether there wasn't more to this case than has ever been revealed – but then perhaps it was just stupidity on a grand scale.

On 30 October 1796, the members of the gang, along with the unfortunate Lesurques, were taken from their prison cells and prepared for execution. The twenty-minute journey from the Conciergerie to the Place de Grève where the guillotining was to take place was the most moving anyone present could remember. As the wagon rolled through the streets Couriol, standing at the front, repeated over and over to the crowd, 'I am guilty, Lesurques is innocent!' People were horrified. Even on the scaffold, moments before the blade silenced him forever, Couriol screamed, 'Lesurques is innocent!'

It made no difference; after hugging his wife and children, a tearful Lesurques went to his death.

Dubosq, the man named by Couriol, was finally captured. He did indeed bear a remarkable likeness to Lesurques. He was eventually tried and executed, four years after Lesurques had answered for the same crime. To this day, despite a general acceptance that he was innocent, Lesurques has never been reprieved.

In many cases victims of a crime might need to be identified too, particularly when that crime is murder. The earlier case of Catherine and John Hayes provides a grisly example of this.

At around dawn on 2 March 1725, a watchman discovered the severed head of a man lying on the muddy foreshore of the Thames at Westminster. It had obviously not been there very long, as decomposition was yet to really set in. The facial features were still intact, meaning that with luck someone might recognise the unfortunate individual. The head was presented to local magistrates, who ordered that it should be cleaned up and its hair combed. After it had been prepared in this way it was taken to St Margaret's Parish Church and stuck on a pole for all to see. The queue to view the remains was apparently so long that traders worked the crowd selling food and water. Parish constables were stationed near the head and around the graveyard, the idea being that the guilty party would surely react in some way if they saw the head. There was also an age-old belief that if a murderer touched the corpse of their victim it would bleed. Therefore anyone who seemed particularly upset at seeing the head was forced by the constables to touch it so that they could observe whether blood oozed forth from it.

Perhaps unsurprisingly, this approach failed to produce a suspect, and it was not long before the head began to decay, and to be pecked at by the local birds. Fearful of it becoming unrecognisable, the magistrates ordered that it be immersed in a large jar of gin to preserve it and that it then be taken inside the church. This was duly done and that, for the time being, was that.

Catherine Hall was a dominant, attractive woman who drew admirers easily. She was born near Birmingham in 1690, the daughter of a pauper, and left home at the age of fifteen to seek her fortune in London. On her way she fell in with several military officers, who took a shine to her and brought her with them to their billets at Ombersley, Worcestershire, where she stayed with them for some time. She eventually left them and was next picked up by a respectable farmer called Hayes. He was much older than her and she quickly formed a relationship with his son John instead. The two were married in secret. When John's father found out, seeing that it was too late to do anything about the relationship, he set his son up in business as a carpenter. However, the rural life wasn't enough for Catherine – she wanted more. She wanted London and all that it had to offer her. After putting considerable pressure on her new husband, she finally convinced him to move there. The pair established a lodging house and soon also became successful coal merchants, money-lenders and pawnbrokers. They quickly amassed considerable savings. Later, Catherine took in two young lodgers called Thomas Wood and Thomas Billings.

An organ-builder's apprentice by the name of Bennet had by now seen the head on display in St Margaret's. Having done so, he felt compelled to call on Catherine at her residence on Tyburn Road (now Oxford Street), to tell her that he believed the head to be that of her husband John, with whom he had once worked. Catherine was incensed. She assured Bennet that John was quite well and warned him that if he continued to spread such nasty false rumours she would have to ask the police to arrest him.

But one Samuel Patrick had also been to see the head, and he too felt certain that he recognised it. Later that day, he told anyone in the Dog and Dial pub who would listen that the head bore a striking resemblance to John Hayes of Tyburn Road.

Thomas Billings, one of Catherine's lodgers, happened to be drinking in the pub at the same time. He assured the company that all was well and that he had left John Hayes sleeping soundly when he set out from home that morning. Despite this reassurance, several of Hayes' friends remained suspicious. Eventually, a man by the name of Ashby went round and asked Catherine about her husband to her face. She came up with a most bizarre explanation, telling him that John had been forced to flee to Portugal, having killed a man during a quarrel. Ashby was quite rightly unconvinced by this explanation, especially since Billings had completely failed to mention this rather dramatic occurrence. Another friend of Hayes, a Mr Longmore, also questioned Catherine on the matter and similarly felt sure that she was not telling the truth. As a result the two went to see a magistrate, who agreed with them that it all seemed rather suspicious and who issued a warrant for Catherine's arrest. She was found in bed with Billings. Both were promptly arrested, as were two other lodgers, Thomas Wood and a Mrs Springate.

Catherine now asked to see the head and was taken to it. On being shown the pickled remains she snatched the jar up in her arms and screamed dramatically, 'Oh, it is my dear husband's head!' and kissed the jar. Clearly this was not a sufficient display of her feelings, for then, in one of the most bizarre incidents in the history of forensic detection, she lifted the now seriously decomposed head itself from the jar by its hair and kissed it passionately on the lips. She then asked for a lock of her dead husband's hair. The constable refused, telling her the head was bloody and that she already had enough blood on her hands. Perhaps realising that her dramatic display had not fooled anyone, Catherine then passed out.

Thomas Wood proved to be the weak link in the group. When questioned he quickly broke and confessed that he and Thomas Billings had both been Catherine's lovers. Tired of her husband's 'mean spirits', Catherine had persuaded the two men to murder him. They got him drunk on six pints of wine so that he fell asleep, at which point Billings hit him over the head with a coal-hatchet. Wood was then handed the hatchet and told to finish the job, thus ensuring that he was fully involved in the murder. He struck John Hayes across the head several times, until they were sure he was dead. They then put his head above a bucket and sawed it from his shoulders using a sharp carving knife. Catherine wanted to boil the head in order to destroy its features, but this was deemed a step too far by Wood and Billings and they refused to do it. Instead they took it away in a bucket and threw it onto the foreshore of the River Thames. They then returned home and dismembered the rest of the body before throwing the bits into a pond in Marylebone. When the pond was dredged, the rest of the body was indeed discovered.

Catherine Hayes was not to be charged with murder but rather 'petty treason' – her husband was supposed to be her lord and master and she had rebelled against him. The penalty for this was not hanging but the far worse fate of being burnt at the stake. On learning this, Catherine finally confessed her part in events while trying to pin the crime on Wood and Billings. However, this made little difference and she was condemned to be burnt.

While being held in prison, Catherine attempted to poison herself, no doubt hoping for a less painful end. The attempt, however, failed, and on 9 May 1726 she was duly burnt alive at Tyburn, where the Marble Arch now stands. It was normal practice to strangle the condemned before the flames reached them, an act of mercy, but in Catherine's case the executioner burnt his hands while throttling her and so was unable to complete the work. She survived the flames for longer than anyone could have imagined. It was said that her screams could be heard all over London. She was the last woman in England to be burnt alive for petty treason (although the bodies of women were burnt after execution until 1790).

As for the amorous, murderous lodgers: Thomas Billings was hanged in chains in Marylebone Fields, close to the pond where he had dumped John Hayes' body. Thomas Wood avoided the gallows by dying of fever in prison.

In this case the correct identity of the victim was established serendipitously by a few people who knew him seeing the severed head while it was on display. However, we can well imagine that without this good luck, the culprits might have escaped punishment for their crime. Equally, if the head had been boiled as Catherine Hayes had wished, then even those who knew him well would almost certainly not have been able to make an identification. Better methods were needed, though their development would take more than a hundred years.

One of the major problems that continued to confront the police was the issue of how to identify habitual thieves. A man arrested in, say, Nottingham, might already be wanted for crimes in London, Liverpool or Norwich, but would escape punishment for them because the authorities had no way of making this connection. Advances in transport infrastructure such as a growing railway system only exacerbated the situation: criminals were now able to move quickly around the country to commit crimes in widely different locations within a short space of time, and with no system to keep track of them, they were at complete liberty to do so.

It was the French scientist Alphonse Bertillon (1853–1914) who took the first major steps towards solving this problem, though he arguably would not have done so were it not for the influence of several important figures. The first of these was the Belgian astronomer, Lambert Quetelet, seen by many as the father of modern statistics. In his 1835 book *Sur L'Homme et le développement de ses facultés* (published in English in 1842 as *A Treatise on Man and the Development of His Faculties*), he attempted to apply the statistical method to the development of human physical and intellectual faculties – in layman's terms, he wanted to know what made people tick. Alphonse's father, Louis-Adolphe, then a young medical student, was fascinated by Quetelet's ideas on 'social physics'. During the 1848 Revolution in France, he found himself in prison for six months with one of his professors, Achille Guillard, who was also interested in this field. (Guillard, who was considered a dangerous liberal, also invented and developed demography, the study of regional groups and races.) The two must have got on well because shortly afterwards Louis-Adolphe married Guillard's daughter Zoe. Guillard and Louis-Adolphe went on to found the School of Anthropology in Paris, and in doing so formalised a new science. There can be little doubt that the indirect influence of Quetelet and Guillard and the direct influence of his father shaped Bertillon's work.

That said, Bertillon did not get off to a promising start in life; he was a rebellious child, often referred to as *l'enfant terrible.* He was thrown out of schools and his German tutor resigned in disgust. The death of his mother in 1866 only caused his behaviour to deteriorate further. As an adult he drifted through life, working first as a teacher in England, then joining the army, and finally settling down as a clerk in the department of the Prefecture of Police (thanks almost solely to his father's influence). The work, however, was repetitive and boring and almost drove Bertillon out of his mind. To alleviate this boredom he began to apply himself to the problem of identification. He quickly realised that most of the techniques being used by the police were at best defective and at worst totally useless – there was no proper system in place to facilitate efficient identification. Bertillon took his inspiration from Quetelet's *Anthropometry, or the Measurement of Different Faculties in Man* (1871), reasoning that if human faculties could be measured and recorded, surely physical characteristics could as well.

He began to work on a system of identification that today we would refer to as 'photo-fit pictures'. He cut up photographs and stuck the pieces to sections of cardboard so that they could be arranged together in different combinations of component parts (ears, eyes, noses, mouths, etc.) to create new faces. Using them, a witness could construct a rough likeness of a person they had seen. A refined version of this system is still used as a method to identify people to this day.

At first Bertillon's system convinced few people but, despite lack of support from his colleagues, he persisted. Thanks to Quetelet's work, and that of his own father, he knew that human characteristics tend to fall into statistical groups. And he knew – as all French hat-makers and tailors had known for years – that no two human beings have all the same measurements. He realised that if he could devise a quick and simple system of measuring various parts of a criminal's anatomy (such as the circumference of their head and the length of their arms, legs and fingers), he could then match these measurements against any individual that came through his door. Through doing so, he would quickly be able to ascertain if a person was giving their real name or, if they did not have a name (as in the case of some dead bodies), identify them and give them one. He then went on to devise an easy-to-use card index in which to store all this personal information. This done, Bertillon was satisfied that he had designed the ideal method of identification.

With high hopes, he submitted a report on his work to his chief, Louis Andrieux. Andrieux completely ignored it. Not one to give up easily, Bertillon submitted another, more detailed than the first. This finally managed to attract Andrieux's attention. He sent for Bertillon, who hurried to his office full of excitement. However, far from heaping deserved praise on Bertillon as expected, Andrieux poured scorn on the idea. Bertillon's attempts at explanation fell on deaf ears; he was thrown out of Andrieux's office and returned to his desk. As if this wasn't enough, Andrieux also wrote an angry letter to his father telling him that he thought his son was 'quite mad'. Luckily, in spite of Bertillon's chequered background, his father did not take such a dim view of things; when Bertillon submitted his paper to him he read it with interest and then told him that he thought it was 'a very important idea'. He went on to say, 'If this works, it will prove what I have spent my life trying to demonstrate . . . that every human being is unique.' He was normally a reserved, unemotional man, but on this occasion Bertillon saw tears in his eyes.

However, even with this support, Bertillon found it hard to

publicly advance his ideas when Andrieux would not allow him to put them into practice. Andrieux was not an intelligent man, and it is probable that he resented his subordinate's abilities and education. Still, Bertillon persisted in taking the measurements of anyone who would allow him to do so. As time progressed he was promoted and eventually Andrieux retired. He was replaced by Jean Camecasse. Although more enlightened than his predecessor, he was still rather dubious about Bertillon's technique. It took over a year of persuasion on the part of Bertillon's father, as well as the intervention of the lawyer Edgar Demange, to finally convince Camecasse to take Bertillon's ideas seriously. In November 1882, he gave Bertillon three months to prove the efficacy of his theories. If, during that time, he managed to identify just one habitual criminal using his methods, Camecasse would allow the experiment to continue. The gauntlet had been thrown down and, even though he knew he did not have long to prove himself, Bertillon rose to the challenge.

The following day, with the help of two clerks appointed by the prefecture, Bertillon began his work. He knew identifying a criminal in the allotted time span wasn't going to be easy, but he was determined to succeed. Over the previous two years he had settled upon a system that involved taking eleven specific measurements from the body of a person. He had estimated that the chances of any two people having exactly the same eleven measurements were more than four million to one. In addition, to each of his identity files Bertillon added two photographs, one of the face front-on and the other in profile (see Plate 1). He also included a *portrait parlé*, a description that involved outlining any distinguishing features that the individual might have, such as tattoos, moles, birthmarks, scars or anything else that might help a police officer identify his suspect. These detailed filing cards were all stored in an eighty-one-drawer cabinet.

Throughout December 1882 and January 1883, Bertillon continued to work tirelessly. As the three-month deadline grew closer though, he started to feel anxious; he even considered asking Camecasse to extend his time limit. By now he had other detractors, including the great French detective Gustave Mace, who thought the experiment a waste of both time and money.

It was towards the end of February that Bertillon finally made his breakthrough. Shortly before he was due to go home one day, a suspect was presented to him who gave his name as Dupont. His face seemed familiar and he was found to have a mole near his left eyebrow. Bertillon set about putting his system into action. He took the necessary measurements and began to flick through his index cards. Those present said later that Bertillon was trembling with anticipation as he searched. After matching the measurements with those on one of his cards, Bertillon declared that the suspect in custody was in fact a man called Martin who had been arrested on 15 December 1882 for stealing bottles. Not only did his measurements match but so did his *portrait parlé*, which mentioned the mole by his eyebrow. Lastly, the photograph that had been taken of Martin at the time of his arrest confirmed that it was the same man. At first Martin denied his true identity, but when confronted with Bertillon's evidence he was forced to admit that he had lied. It was a triumph for Bertillon and, indeed, for his father, who alas died a few days after this success. He had, however, lived to see his son's system and his own life's work vindicated.

Alphonse Bertillon, whose method of identifying criminals using anthropometric measurements revolutionised criminal detection.

Over the next few months, Bertillon continued to successfully identify further suspects. It was clear that his system worked. Eventually even Gustave Mace had to admit that Bertillon had single-handedly brought about the greatest advance in law enforcement of the nineteenth century. Within a few years the word *bertillonage* had not only passed into the French language, but also into many others.

Bertillon went on to apply his technique to identify the dead as well as the living. An inspector asked Bertillon to identify the body of a person who had been shot and dumped in the river. The body had been in the water for at least two months before being recovered and was consequently in extremely poor condition, with no remaining features on the face by which it might have been identified. The inspector considered that Bertillon was his last hope, but felt that in this case even his chances of success were slim. However, Bertillon went through his normal procedures, taking measurements from the corpse and referring to his card index. He managed to match at least five measurements and, to the inspector's amazement, discovered that the man had been convicted of a violent assault a year earlier. With the identity of the body established, the inspector was able to pick up the trail of the murderer and made an arrest soon afterwards.

In light of the success of Bertillon's methods, in 1888 a new Department of Judicial Identity was established at the prefecture. Bertillon was made its first head. He had come a long way, but he still had further to go – in 1892 he became involved in the case that was to really establish him as a household name in France. It involved the notorious anarchist Ravachol, one of the most famous criminals in the country at the time.

Ravachol – real name François Claudius Koenigstein – was born in 1859 at Saint-Chamond in the Loire, the son of a Dutch father (Jean Adam Koenigstein) and a French mother (Marie Ravachol). He adopted his mother's maiden name after his father abandoned the family when François was only eight years old, leaving him to support his mother, sister, brother and even nephew. For a while he worked as a dyer's assistant, but he didn't keep the job for long and subsequently picked up what money he could playing the accordion at society balls.

Wandering through France looking for work (always being paid a pittance when he finally found some) taught Ravachol to hate capitalism. At eighteen he began to read Eugène Sue's *Le Juif Errant* (*The Wandering Jew*) and started attending a collectivist circle. As a result he became a convinced atheist, socialist and anarchist. As well as *Errant*, he was strongly influenced by Pierre Proudhon, Michael Bakunin and Prince Peter Kropotkin. Kropotkin argued that zoological evidence indicated that animals live by 'mutual aid' and said that if humans could rid themselves of all their law-makers, judges, police officers and MPs, then they could live the same way. Proudhon believed in a stateless society where people would live through goodwill and reason.

On May Day 1891, an anarchist demonstration at Clichy was broken up by the police. Its leaders were arrested and badly beaten. Two of them were sentenced to long terms of imprisonment. Six months after this, however, the home of the advocate general, Léon Bulot, who had been the presiding judge at the trial, was blown up by a bomb. Not long afterwards, the same thing happened to the home of Benoît, the prosecuting counsel, who had tried to get the death sentence passed on the anarchists.

Both the police and the government's security services began searching for the culprits.

Someone tipped off a government spy, giving him the name Chaumartin in connection with the affair. Enquiries quickly established that Chaumartin was a technical schoolteacher in St Denis. He was arrested and 'interviewed'. Under this interrogation he finally admitted that, although he had planned the bombings, they had been carried out by a fanatical anarchist called Léger. The authorities quickly discovered that Léger was in fact a known revolutionary – none other than Ravachol. In 1891 Ravachol had been arrested for the murder of an old man and his housekeeper in the Forez Mountains. However, he had escaped and gone on the run. Later that year, two elderly ladies who ran a hardware store in St Etienne had been murdered with a hammer during the course of a robbery. The description of the killer matched Ravachol perfectly.

Ravachol was eventually arrested at Restaurant Véry on the Boulevard Magenta in Paris, thanks to an observant waiter who noticed a scar on his left hand. He remembered that this scar had been mentioned as part of the description of Ravachol that authorities had given out and he informed the police. Ravachol fought like a wild man when they tried to arrest him and had to be subdued with a considerable amount of brutality (see Plate 2). He was taken to the prefecture, where Bertillon noted his measurements. He refrained from taking Ravachol's photograph at the time, as his face was so swollen from the beating he had received. A few days later, however, he did manage to photograph him. Much to everyone's surprise, Ravachol sat quietly for him. Bertillon later sent him a framed copy of the photograph for which Ravachol was grateful, commenting, 'That Bertillon is a gentleman.'

Gentleman or not, Bertillon quickly identified Ravachol and established that he had been arrested previously for smuggling and burglary under his old name, Koenigstein. This was an enormously important connection to have made – it almost certainly meant he was the same man the police were looking for in connection with the murder of the old man and the two shopkeepers, as well as for other offences such as forcing his way into graveyard vaults. He was sentenced to life imprisonment for his anarchist activities at his first trial, but during his second trial was found guilty of murder and grave-robbing and sentenced to death. Much to the dismay of the anarchist movement, he eventually confessed to his crimes. As a result, Ravachol was denounced as an *opéra-bouffe* revolutionary (fit only for a comic opera) by, among others, the anarchist Kropotkin. Any sympathy the public might have had with him quickly evaporated and he went to the guillotine screaming, 'Goodbye, you pigs, long live anarchy!' It is a tragic postscript to the story that – before the anarchist movement had become disenchanted with Ravachol – a fellow anarchist had shown support by bombing the restaurant where he had been arrested, murdering the proprietor and a customer.

As a result of the crucial part his methods played in the case, Bertillon was now a household name in France, the 'Sherlock Holmes' of Paris. Indeed, he is referenced in the Sherlock Holmes story 'The Hound of the Baskervilles' – a client refers to Holmes as the 'second highest expert in Europe' after Bertillon. He crops up again in 'The Naval Treaty', where we are told that Holmes himself 'expressed his enthusiastic admiration of the French savant'. There is no doubt that a great many cases that would otherwise not have been solved owe their successful conclusion to

Bertillon's system of measurement. However, things were changing and a new system would soon come to take centre stage.

People have been aware of the patterns we all have on the tips of our fingers for thousands of years. Examples of fingerprints have been found on the walls of Egyptian tombs, and as decorative motifs on ancient pottery from various cultures. Perhaps more surprisingly, it seems that there was a crude sense that fingerprints were in some way representative of a person's individuality; in ancient Babylon in the second millennium BC, fingerprints were sometimes used in order to seal a legal contract. Later, in China around AD 300, handprints were used in evidence in a trial for theft, while in AD 650, Kia Kung-Yen, a Chinese historian, remarked upon the fact that fingerprints could be used as a form of authentication.

But while this consensus that fingerprints had a certain uniqueness about them persisted, it was many hundreds of years before this would be scientifically described or studied.

In 1684 the renowned English botanist Nehemiah Grew (1641–1712) published a paper describing the ridge structure on the skin covering a person's fingers and palm. Almost a century later the German anatomist Johann Mayer (1747–1801) stated for the first time outright that no two prints were exactly alike; that in fact all were completely unique. This was obviously of enormous theoretical importance for forensic science, though it would be a little while longer until such knowledge was put to practical use. It was the British civil servant Sir William Herschel (1833–1917) who seems to have been the first to use fingerprinting in a really formalised system (see Plate 3). He used fingerprints when paying pensions to Bengali soldiers to stop impostors from being able to collect money. Each of the soldiers had to register their fingerprints on their pay books and also provide fingerprints when collecting their pension. Any impostor would quickly be revealed when his prints did not match up with those in the pay book. This system apparently worked extremely well, but the Bengali inspector general of prisons nonetheless rejected Herschel's idea of creating a larger system of fingerprint classification and analysis. Herschel returned to England in 1879.

At about the same time, a Scottish surgeon, Dr Henry Faulds (1843–1930), was working in Japan, teaching physiology to medical students at the Tsukiji Hospital in Tokyo. During his time there he happened to notice the marks of fingerprints visible on some Japanese pottery. He became interested in the various differences between them, and began to study the distinctive 'whorls' on the fingerprints (which are also known as papillary lines). Several years later, this purely academic work was to have a very worthwhile application. In 1879, while investigating a burglary in Tokyo, the Japanese police recovered a set of grubby fingerprints on a whitewashed wall. A man was later arrested on suspicion of the crime, but vehemently protested his innocence. The police had heard of Faulds and his interest in fingerprints, so they approached him for help. Faulds took the suspect's fingerprints and compared them with those discovered at the scene. It was quickly apparent that the two sets of prints were entirely different and as a result the man was released. A few days later another suspect was arrested; this time the prints did match and the culprit quickly confessed to the crime.

Faulds published his first paper on the subject of fingerprints in the scientific journal *Nature*. In it he discussed their usefulness in establishing identity and proposed the method of recording them in ink. When Herschel returned from India and heard about Faulds' work he was convinced that 'his' discovery had been stolen. Strong letters were exchanged through *Nature*. In reality both men independently did their bit to advance fingerprinting (or dactyloscopy, to give it its proper name) as a method of identification.

When Faulds later returned to the UK from Japan in 1886, he explained his ideas to the Metropolitan Police. They were dismissed. He then wrote to just about anyone he thought would listen, including Charles Darwin. Although Darwin was interested, he felt he was too old and ill to get involved in the matter himself. Instead, he passed the information to his cousin, Francis Galton, who was interested in anthropology. Galton was a sportsman, explorer, meteorologist and psychologist. He was also a believer in Bertillon's system of identification. Not only had he given a lecture on *bertillonage* to the Royal Institution, but he had also visited Bertillon himself in Paris. Although Bertillon's system impressed him, he found it too complicated. He saw the potential of fingerprints as an easier method of identification, but did not yet properly involve himself in the emerging field – he simply forwarded Faulds' communication to the Anthropological Society of London. When he returned to the topic some years later, Galton, having heard of William Herschel's reputation in the field, made contact with him rather than Faulds. Galton and Herschel got on well, as a result of which Herschel handed over all his material to Galton, who set about establishing fingerprints as the major system of identification in forensic terms.

He needed to develop a proper system of classification. He

knew that it was essential for any such system to be simple – previous attempts at clarification had been very complicated and this was certainly one of the reasons that the authorities remained dubious about putting fingerprinting into practice. Galton began to observe recurring shapes and configurations of lines and that most fingerprints are centred around a 'triangle' where the ridges run together. These triangles are called deltas and fall into four basic patterns: no triangle, triangle on the left, triangle on the right, more than one triangle. In 1891, Galton published a paper discussing his findings on fingerprints in *Nature*. In it, much to Faulds' fury, he acknowledged his debt to Herschel but made no mention of Faulds. The following year he published his first book on the subject, *Finger Prints*. In it he demonstrated that the chance of a 'false positive' (two different individuals having the same fingerprints) was about 1 in 64 billion. It was an extraordinary piece of work and influenced the then home secretary (later prime minister) Herbert Asquith, who was at the time considering introducing the *bertillonage* system to Britain.

As a result of reading Galton's book, Asquith established a committee to examine both systems in detail. He appointed a Home Office official named Charles Edward Troup to head the inquiry, supported by Major Arthur Griffiths (famous for his book *Mysteries of Police and Crime*) and Sir Melville Macnaghten, who was to become an assistant commissioner of the London Metropolitan Police. Although they liked the idea of fingerprinting because of the system's simplicity, they were also concerned because Galton had still not managed to distil everything he had observed into a fast and accurate practical system. The committee also travelled to Paris and were entertained by Bertillon, whose system they were convinced by, but found complicated. In typical English fashion, they couldn't make up their minds. As they pondered, other countries were already deciding the same question for themselves. Austria, under the guidance of the father of criminology, Hans Gross, went for Bertillon's system, as did Germany. Eventually, in a classic British compromise, the committee decided to introduce both *bertillonage* and fingerprinting.

Meanwhile, in Argentina, a police officer named Juan Vucetich from Dalmatia (a region of Croatia) was to be responsible for a first in the history of forensics. Vucetich was an energetic man and, in 1891, having resided in Argentina for seven years, was made head of the Statistical Bureau of the La Plata Police. He and his team were ordered to introduce the *bertillonage* system and so set about measuring people and recording their statistics. During this time, however, Vucetich read about Galton's work on fingerprints in the *Revue Scientifique*. The article, written by H. de Varigny, praised the concept of fingerprint identification, but also pointed out that – despite Galton's success – he had still not fully solved the problem of classification.

Vucetich was intrigued by this idea and decided to take up the challenge. He, too, quickly understood that the essential features of fingerprints were their triangles or deltas, and that there were four basic types: those with no triangle, those with a triangle on the right, those with a triangle on the left, and finally those with two triangles. He numbered these types one, two, three and four when referring to an individual's fingers, and assigned the letters A, B, C and D when referring to thumbs. So, for example, a suspect might now have their fingerprints recorded as: B, three, three, four, two. The system was easy to store and arrange, making it simple to crosscheck for matches.

In an echo of Bertillon's situation years before, Vucetich unfortunately found that his bosses did not share his enthusiasm for fingerprints. But once again the fates were to intervene. In June 1892 a double murder was committed in the small coastal town of Necochea, not far from Buenos Aires. The victims of the crime were two young children, a girl of four and a boy of six. They had been bludgeoned to death. Their mother, a twenty-six-year-old unmarried woman by the name of Francisca Rojas, had not only discovered the bodies, but also claimed to have seen a man running from the scene of the crime. The man, she stated, was her lover, a farm worker called Velásquez. She said he had become a nuisance, making threats against her and her children in order to force her to marry him. When she came home he had apparently run past her and out of the house, after which she found her children dead in a bloodstained bed.

Velásquez was arrested and interviewed at length. This almost certainly involved a certain degree of torture. However, in spite of this, he continued to protest his innocence. Other 'medieval' tricks were played on him, such as tying him up and leaving him on the bed with the murdered children all night. He still denied any involvement. Given all Velásquez had endured, some doubts began to be expressed about his guilt at this stage, but it was decided to try torture for another week. Even after suffering serious injuries, though, he continued to proclaim his innocence.

Suspicion returned to the children's mother, Francisca Rojas. It was discovered that she had a young lover who had allegedly said that he would not marry her because of her 'illegitimate brats'. Alvarez, an investigating officer, now arrested her and tried similar rather questionable techniques to those he had used on Velásquez. Hoping to terrify her into confession, he had her tied up and left outside her own front door so that the spirits of the two children could take their revenge. He even had men make angry noises outside to try to convince her they were coming to collect her evil soul.

At last, when all these techniques had failed, he did what he should have done in the first place and searched the murder scene. It didn't take long before he discovered a bloody mark on a door. Examining it more closely, he realised that it was a fingerprint, and a good one at that. He cut the plank bearing it from the door and took it back to the police station. He then took the prints of Francisca Rojas and compared the two. They matched. He asked Rojas if she had touched her children at all after she had found them dead. She said she hadn't. If that was the case, he asked, how did her bloody thumbprint get onto the door? He showed it to her. Confronted with this evidence, Rojas finally confessed to murdering her two children with a rock so that she would be free to marry her young lover. She was convicted and sentenced to life imprisonment. The case is generally acknowledged as the first time that a fingerprint was used to solve a murder.

The Rojas case did for Vucetich what the Ravachol case had done for Bertillon, and he became the most celebrated detective in Argentina. In 1896, Argentina adopted fingerprinting as its main system of identification, and by the first decade of the twentieth century, every major country in South America had followed suit. In England, Galton continued to struggle to devise a satisfactory classification system, but help was about to appear from an unexpected source.

A civil servant named Edward Richard Henry was the inspector general of police in Nepal. In 1891 he introduced

Bertillon's system there, amending it to involve six measurements rather than eleven in order to make it simpler and faster to use. Even then, he still found the system too complicated, as well as vulnerable to the enthusiasm of the clerks taking the measurements, who often couldn't see the difference when it came to 'only a few centimetres', and frequently got them wrong.

Whilst on leave in England, Henry visited Galton. The two men got on well, and when Henry returned to Calcutta he took with him all Galton's notes. Henry also saw how difficult the process of categorisation was going to be. However, during a train journey in 1896, he suddenly realised, quite out of the blue, how the deltas (those triangular shapes found on the tips of the fingers) could be used to create a proper system of identification. They fell into several clear types. Henry observed that 'These deltas may be formed by either the bifurcation of a single ridge or by the abrupt divergence of the two ridges that had hitherto run side by side', and additionally their triangular shape conveniently lent itself to geometric measurement. He realised that all he had to do was establish the limits of the triangle or 'the outer and inner terminus'. A line could be drawn between these two termini and the number of papillary lines that this line intersected then counted with a needle. This number was the core of Henry's classification. The vast majority of fingerprints fall into the simple loop and delta system. There were occasional examples of what he termed 'accidentals' (those prints that for one reason or another didn't match any of the types), but fortunately these could still be absorbed into the general system, meaning that he now had a practical means of categorising any fingerprint. By 1897, fingerprints had become the sole means of criminal identification in India. And by 1902 they were proving three times more successful at identifying criminals than Bertillon's system.

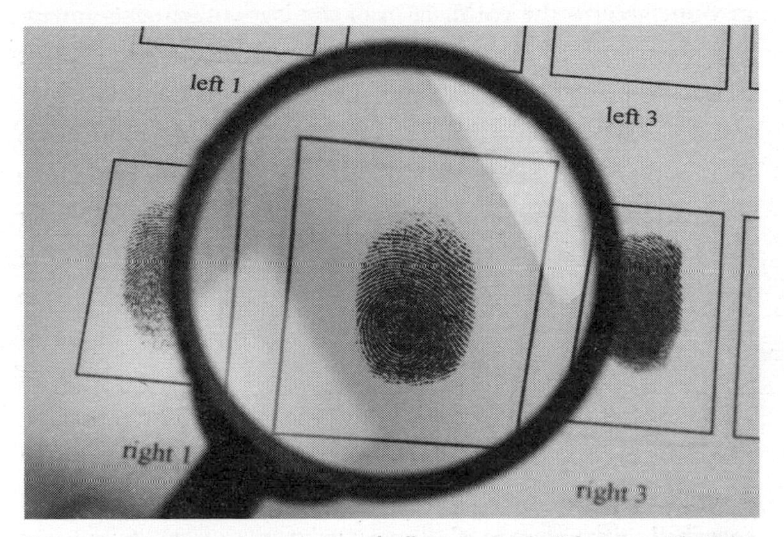

Fingerprints stored in modern police records, illustrating a 'whorl' fingerprint. The Henry system described three basic fingerprint patterns – loop, whorl, and arch – which together constitute the majority of fingerprint variations.

Considering that fingerprinting stood to supplant his own system, it is perhaps surprising – and to his credit – that Bertillon readily accepted its value. Indeed, since 1900 he had been adding fingerprints to his own files, which proved to be invaluable. On 17 October 1902, Bertillon was asked to attend the scene of a murder at the Rue du Faubourg Saint-Honoré. The victim was a valet by the name of Joseph Reibel. He was discovered sitting in a chair. His shirt-tails were pulled out of his trousers and his legs were outstretched. The murderer had strangled him by hand. The room was in a mess with overturned furniture, suggesting that a struggle had taken place. This, combined with the discovery that some drawers and a cabinet had been forced open, suggested that the motive was robbery. Bertillon, however, wasn't convinced; the amount of money that had been stolen was hardly enough to justify murdering a man.

A glass panel in the cabinet had been smashed and there was blood on the glass, suggesting that the culprit might have been injured. One of the police inspectors went to pick up one of the shards, but Bertillon stopped him; he had spotted a fingerprint. In fact, it transpired that the killer had left an almost perfect set of fingerprints in blood. With great care, Bertillon had the glass taken back to his laboratory and photographed it. From this he produced a first-class image of three fingerprints and a thumbprint.

Of course, Bertillon now wanted to make a match, but the only chance of that was if the killer already had a record. Initial results were discouraging, but Bertillon kept searching and eventually, in a moment of great elation similar to the one he'd experienced during the Dupont case, he discovered a match for the fingerprints. The card belonged to a well-known twenty-five-year-old swindler called Henri-Léon Scheffer. Scheffer was tracked down to Marseilles but before the police had had time to arrest him, he had handed himself in and confessed to the murder, explaining that he and Reibel had been lovers and had fought, and that he had stolen the money to try to cover his tracks. Once again fingerprints had triumphed. Bertillon could now add to his long list of successes being the first man in Europe to solve a murder by way of fingerprints. Even so, he still did not adopt fingerprints as his primary means of criminal identification, refusing to give up his own system of body measurements. To have accepted that fingerprints alone could operate as well as, or better than his own system, rather than simply being a helpful addition to it, would have meant admitting that his life's work no longer had a useful application.

It was also around the turn of the century that fingerprints at last began to show their worth in Great Britain. On Derby Day 1902, Scotland Yard deployed a number of fingerprint experts to the horse race. Since its beginnings in 1780, the Derby had become a well-known target for pickpockets and other criminals, who would flock there from across the country. All day long the fingerprint experts inked the fingers of arrested suspects. When, at the end of the day, they came to crosscheck the fingerprints against those on file, they discovered that twenty-nine out of the fifty-four men from whom they had obtained prints had previous convictions. When they were put before the magistrate the following day, these records were produced. With such strong evidence of their previous convictions, they were all sent to prison for at least twice as long as a first-time offender would have been.

But it wasn't until 1905 that fingerprints were first used to solve a murder case in Britain. The case in question was that of the infamous Deptford murders.

At 8.30 A.M. on Monday 27 March 1905, one William Jones arrived at his place of work, Chapman's Oil and Colour Shop on the High Street, Deptford. He found it still closed which immediately concerned him, as it would normally be opened up by the owner, seventy-one-year-old Thomas Farrow who lived in the small flat above it with his wife Ann, aged sixty-five. Jones tried knocking but received no response. Becoming even more concerned, he peered through a window and saw that several chairs had been overturned.

Now seriously worried, he ran to a local store where he found employee Louis Kidman; Jones asked him to come back with him to force an entry. Once inside, they discovered Mr Farrow lying on his back in the downstairs parlour. He was dead. They then discovered Mrs Farrow in bed in the upstairs flat, still alive but badly injured. Both showed signs of a serious and sustained beating. The police and a doctor were summoned, and Mrs Farrow was rushed to hospital.

An empty cash box was found on the floor, which would usually have contained the weekly takings. Jones explained that Mr Farrow would normally pay these in to the bank on a Monday morning. Trying to be helpful and clear the scene, Sergeant Albert Atkinson pushed the cash box aside with his bare hands. When they became aware of this, Chief Inspector Frederick Fox and Assistant Commissioner Melville MacNaghten (head of the Criminal Investigation Department) decided to take over on the case and preserve any remaining evidence.

The motive then had been robbery. The police were able to deduce several other details. From where they had been found and from the evidence of the scene, it was clear that Mr and Mrs Farrow had been attacked separately. Both were still in their nightwear, and there was no sign of a forced entry, so it was likely that Mr Farrow had opened the door to his attackers before being beaten unconscious. The assailants must then have gone upstairs to the flat and attacked Mrs Farrow, before finding the cash box and stealing the cash. From the blood trails at the scene, it looked as if Mr Farrow had then somehow managed to get up again, only to be beaten once more, this time to death. The discovery of two black stocking masks pointed to the likely involvement of two perpetrators, who had coolly washed their hands in the sink after killing the shopkeeper.

MacNaghten looked closely at the empty cash box and saw what appeared to be a fingerprint on the inside tray. As a member of the Belper Committee established to assess methods of identification, which had recommended the use of fingerprints five years previously, he wondered whether this might be a good opportunity to test the new technique. He used a handkerchief to carefully pick the cash box up, before having it wrapped in paper and taken to the fledgling Fingerprinting Bureau at Scotland Yard.

The bureau was headed by Detective Inspector Charles Stockley Collins, who was by then regarded as the foremost English fingerprint expert of his time. Despite the earlier successes of the method, especially in identifying previously convicted criminals who tried to pass themselves off under pseudonyms, the technique was still considered unwieldy. The police knew that they were risking public ridicule if it failed now, due to the intense scrutiny that a murder case would generate. Furthermore, even if they succeeded in identifying the owner of the fingerprint and charging him, they would still need to convince a jury to convict on the basis of this unfamiliar form of evidence.

Collins examined the print thoroughly and determined that it was made through perspiration and appeared to have been left by the thumb, probably from the right hand. He compared it with those of the Farrows and of Detective Sergeant Atkinson and was satisfied that it did not belong to either of them. Although the Bureau had some 80–90,000 sets of prints on file, there was unfortunately no match among them. This left the police with the daunting prospect of having to find a suspect to compare the print with. Initially they hoped that Mrs Farrow would be able to give a description of her assailants when she regained consciousness. However, this was not to be and tragically she died in hospital on 31 March without saying a word – a serious setback to the investigation.

The police then had to resort to the usual practice of interviewing potential witnesses. Fortunately there was no shortage of them. Several had seen two men running from the scene of the murder at about 7.30 A.M. One of them was described as being dressed in a dark brown suit and cap, the other in a dark blue serge suit and bowler hat. Two of these witnesses, a professional boxer named Henry John Littlefield and a local girl named Ellen Stanton, positively identified the man in the dark brown suit as one Alfred Stratton.

Although he did not have a criminal record, Alfred Stratton was familiar to the police as a 'vagabond' and was known to have contacts in the criminal underworld. His brother Albert was also known to them, and the description of the man in the bowler hat matched him. The identification of Alfred was apparently confirmed when his girlfriend Annie Cromarty told the police that he had disposed of his dark brown coat and changed his shoes the day after the murder; she also recalled him asking for a pair of old stockings. A tip from Cromarty also led police to recover £4 that was buried near a local waterworks. Based on Cromarty's information, warrants for the arrest of both the brothers were issued. They were taken into custody on 2 April and, while being held, had their fingerprints taken. When Detective Inspector Collins received the two sets, he compared them to the print on the cash box. He concluded that the print matched Alfred Stratton's right thumbprint. The brothers were charged with murder and the trial was set for 5 May at the Old Bailey.

MacNaghten, Collins and Richard Muir, the prosecutor for the Crown, knew that they would face an uphill battle. Since the fingerprint was the only tangible evidence that they had, the case would stand or fall on whether it convinced the jury, and the defence would try their best to undermine it. Even fingerprinting pioneer Henry Faulds was a vocal detractor, because he had the mistaken notion that a single fingerprint match was unreliable. The defence therefore retained him as a witness. Also set to testify for the defence was Dr John George Garson, who advocated anthropometry (the English term for *bertillonage*) over fingerprinting as a means of identification. Both men were professional rivals of Edward Henry, the commissioner of the Metropolitan Police who had established the Fingerprint Bureau and who was responsible for the acceptance of fingerprinting into the British legal system; he was also in attendance.

The prosecution called over forty witnesses to the stand, since Muir and his team wanted to place the two defendants at the scene of the crime. Despite Muir's inherent distrust of eyewitness testimony, he was counting on the consistency of these witnesses to reinforce the evidence of the fingerprint. Although some of them, such as Henry Alfred Jennings, a local milkman, were not able to make a definite identification of the defendants (but were consistent in describing their general appearance), others, such as Henry Littlefield and Ellen Stanton, were positive in their identification of Alfred Stratton. The Home Office pathologist who did the postmortem on the Farrows told the court that the injuries on the couple were consistent with being inflicted by weapons similar to the tools that the brothers had in their possession.

Kate Wade, Albert Stratton's girlfriend, testified that Albert was not with her on the night of the murder, and that he usually stayed with her. Annie Cromarty testified that Alfred had come home on the morning of 27 March with a large amount of money, without explaining where he had obtained it. She added that he threw out the clothes that he had been wearing that day when he saw the newspaper accounts of the murder, and that he asked her to tell the police, or anyone else who asked, that he was with her on the night of the murder.

However, the defence counsels, H. G. Rooth, Curtis Bennett and Harold Morris, were able to give plausible alternative explanations for events that cast doubt on the prosecution's witnesses. They clearly felt they had done a good job, since they were then confident enough to have Alfred Stratton take the stand. He testified that at about 2.30 A.M. on 27 March, he was awoken by his brother tapping on the window. When he opened it, Albert asked if he could lend him some money for a night's lodging. He replied that he would check if he had some and then went inside to do so. When he came back, Albert was gone. He went out and found his brother some distance away, on Regent Street. It was there that several witnesses had seen them. Alfred told his brother that he had no money but offered to let him stay for the night. Albert agreed and slept on the floor, and the brothers stayed together until nine in the morning, after which Albert left. Alfred went on to explain that the £4 the police had recovered he had won boxing. He had, he said, buried it three weeks prior to the murders for safekeeping, and had been intending to give it to Annie Cromarty.

Before calling Inspector Collins to give his evidence about the thumbprint, Muir called William Gittings, who worked in the jail where the Stratton brothers had been confined while awaiting trial. Gittings explained that during a conversation with him Albert Stratton had said, 'I reckon he [Alfred] will get strung up and I shall get about ten years . . . He has led me into this.' Muir hoped to impress the jury into thinking that this statement could be counted as a confession. He then called Inspector Collins to the stand.

Muir's plan was to first establish Collins' credentials as an expert in the field of fingerprinting and then get him to explain, in layman's terms, how fingerprinting worked as a means of identification. Collins was then asked to talk specifically about the fingerprint involved in the case. He showed the jury the cash box that was recovered from the scene and the fingerprint that he was able to obtain from it. He then went on to show how it matched Alfred Stratton's right thumb, pointing out that the print had as many as twelve points of agreement. At the request of a member of the jury, Collins also demonstrated the difference in prints caused by various levels of pressure.

After Collins had given his evidence, the defence called Dr John Garson to the stand. They were hoping to discredit Collins' testimony by establishing Garson's credentials as one of Collins' mentors, thus giving the jury the impression that he was more experienced in the study of fingerprinting. As expected, Garson testified that he could say with certainty that the print taken from the cash box and Alfred Stratton's prints did not match.

However, it was easy for Muir to establish that Garson was not an expert in fingerprinting but in anthropometry, its rival form of identification. Garson had, in fact, spoken out against fingerprinting to the Belper Committee. Muir then dropped a bombshell during his cross-examination of Garson. He called into evidence two letters written by Garson, one to the Director of Public Prosecutions, and the other to the solicitor for the defence. Each said that Garson would be willing to testify for either side in the trial, depending on who would pay him more. In a stroke this rendered his evidence completely worthless. Annoyed by this revelation, the judge commented that Garson was an 'absolutely untrustworthy witness'. Having had the credibility of Dr Garson as a witness shattered in this way, the defence decided not to call Faulds to the stand, fearing that Muir would find some way to discredit him as well. After each side had given their summations, it took the jury just over two hours of deliberation to find the Stratton brothers guilty of murder. They were sentenced to death by hanging, and were put to death on 23 May 1905.

The history of identification – which I will continue to allude to throughout this book – is a history of uniqueness. Proven systems of identification such as *bertillonage* or fingerprinting are able to work because we are all completely individual, something which is enormously useful for the purposes of criminal investigation. The techniques we have looked at in this chapter represent the first successful attempts to integrate forensic methods into justice systems. They demonstrate that police work is made far easier when suspects can be quickly and efficiently tied to (or eliminated from) an investigation. That said, such proofs of identity, however strong, are often only one part of the puzzle – a case constructed using several different forensic techniques in conjunction will build an even more comprehensive picture of events.

2

Ballistics

Only the monstrous anger of the guns.

Wilfred Owen, 'Anthem for Doomed Youth' (1917)

Police Constable George Gutteridge was born in Downham Market, Norfolk, in 1891. He joined Essex County Constabulary in April 1910 and served as constable 489 for eight years before resigning in April 1918 to join the army. He served in France for ten months with the Machine Gun Corps, enduring all the horrors of trench warfare. He then returned to work for Essex Constabulary. He lived with his wife Rose and their two children, Muriel and Alfred, in the small, pretty village of Stapleford Abbotts, working four beats for the Epping division.

On 26 September 1927, Gutteridge was working a split shift. He returned home from duty at 6 P.M. and spent the evening in with his family. He then left home to resume duty at 11 P.M., going to meet his opposite number, Constable Sydney Taylor, who was stationed in the hamlet of Lambourne End. The two met as planned at a conference point on the B175 road running from Romford to Chipping Ongar. Gutteridge departed at 3.05 A.M. to start the mile-long walk home. He never made it.

The following morning, at about 6 A.M., the local postman William Alec Ward was on his rounds when he dropped some mail off at the post office in the little village of Stapleford Abbotts. He then continued along the Ongar road, over Pinchback Bridge towards the village of Stapleford Tawney. It was while he was negotiating a bend that he noticed a large object at the roadside ahead and, as he drew closer, realised that it was the body of a man. The body was slumped against the grass bank in a semi-sitting position, with legs extended out into the road. To his horror, Ward recognised the body as PC Gutteridge. Jumping back into his van, Ward raced to a nearby cottage to summon help before driving to nearby Stapleford Tawney to telephone the Romford police.

The first officer on the scene was Police Constable Albert Blockson, who took charge until Detective Inspector John Crockford arrived from Romford at about 7.45 A.M. The inspector examined the body. Gutteridge was still grasping a pencil stub, while his notebook lay in the road nearby. His truncheon was still in the pocket in which it was usually kept, as was his torch. On the left side of the face, just in front of the ear, there were two holes that appeared consistent with the entry of two large bullets. On the right side of the neck there were two exit wounds. Two further bullets appeared to have been discharged, one into each eye. It was thought that the reason for this might have been the superstition that the last thing a person sees before he dies is photographically imprinted on the retinas of the eyes – the shots had been fired in order to destroy any such 'image'.

The assessment that four bullets had been fired was confirmed when two .45 bullets were prised out of the road surface and two more were recovered from the body during the subsequent postmortem. The time of death was estimated to be about four or five hours prior to the discovery of the body. Because he had been holding his notepad and pencil when found, it was deduced that Gutteridge had stopped a car and had been about to record details when he was shot. The bullets and the cartridge case were handed over to the foremost ballistics expert of his day, Robert Churchill, for examination. Although deformed, the bullets retained sufficient rifling characteristics for Churchill to establish that they had been fired from a Webley revolver.

A full-scale hunt for the killer or killers began. From the outset the murder was connected with the theft of a Morris Cowley car, registration number TW6120, which had been stolen on the same night from the garage of Dr Edward Lovell in Billericay (about ten miles away from the scene of the crime). Neighbours remembered the sound of a car being driven off at high speed during the early hours of the morning. By the time the car had been reported stolen later that morning, however, it had already been found – in Brixton, South London. Its nearside mudguard had been damaged, and blood traces were discovered on the bodywork. The car's milometer also showed that it had been driven forty-two miles – the precise distance of a direct journey from Dr Lovell's garage to Brixton.

The police searched the car and discovered a cartridge case inside. It appeared to have been scarred by some kind of fault in the breech block of the gun which had fired it. This mark resembled a jockey's cap; because of the significance it would later have in the case, the crime became known as The Jockey Cap Murder. The cartridge also carried the letters RLIV, which signified that it was of an old Mark IV type and had been made in Woolwich Arsenal at the Royal Observatory for use in the First World War.

The murder hunt went on for four gruelling months – at one point DCI Berrett and his assistant Sergeant Harris worked for 130 out of 160 consecutive hours. They came to suspect two car thieves, Frederick Browne (46) and Pat Kennedy (42), but lacking any evidence against them were unable to take things further. Two Webley revolvers were found in the River Thames, but Churchill proved that neither of them could have been the murder weapon because neither made the tell-tale jockey-cap mark on a cartridge case.

However, the fortunes of the investigation changed when, on 20 January 1928, police arrested Frederick Browne at his garage near Clapham Junction on suspicion of stealing a Vauxhall car. Browne had convictions for insurance fraud, stealing cars, violence and, most significantly, for carrying firearms. He was searched and twelve .45 cartridges were discovered in his back pocket. After this a further search was made of his car and a fully loaded Webley revolver was found inside the driver's door. More police officers were now called in. As a result, a further sixteen .45 cartridges were discovered wrapped in paper in the office of the garage. Twenty-three .22 cartridges were found as well, in addition to a small revolver. Finally, a search of Browne's rooms in Lavender Hill turned up a fully loaded Smith & Wesson.

Browne was known to have employed Kennedy, a notorious alcoholic, though had sacked him on 17 December 1927 as his drinking problem was affecting his work. Kennedy was an unusual mix, having been born in Scotland to Irish parents – he had an Irish accent but considered himself Scottish. He had been dishonourably discharged from the army for desertion and had convictions for housebreaking, indecency, drunkenness and theft.

When they parted on 17 December, Kennedy took a train to Liverpool. Three weeks later he returned to London in order to get married. Having been away in Liverpool, he had not heard about Browne's arrest. As a result he visited Browne's garage in the early afternoon of Saturday 21 January. He found it locked. When he looked through the crack between the doors he noticed two men in trench coats and hats. He suspected at once that they were detectives and that they were waiting for him. Being careful to ensure he wasn't being followed, he returned home and collected his wife before catching the midnight train to Liverpool.

However, Kennedy could not evade the police for long. At 11.40 P.M. on Wednesday 25 January, he noticed several men in the street outside his lodgings. Realising by the way they were dressed that they were probably detectives, he made a run for it. He left in such a hurry that he didn't have time to put on a shirt or to fasten his trousers or shoes. However, he did not go unnoticed. DS Bill Mattinson, who had arrested him on several occasions and knew him well, approached him. Kennedy pulled a pistol from his pocket and shouted at Mattinson, 'Stand back, Bill – or I'll shoot you!' True to his words, he then pulled the trigger, but fortunately for Mattinson the gun failed to go off (it was later discovered that he had left the safety catch on). Mattinson then grabbed Kennedy's gun arm in one hand and punched him hard with the other, calling for assistance as he did so. Three of his colleagues soon arrived and they were able to disarm and restrain Kennedy.

By the following day, Kennedy was back in London and under arrest in New Scotland Yard. While he was there, Detective Inspector Berrett interviewed him about the murder of PC Gutteridge. Kennedy asked if he could have a few moments to consider his situation. When Berrett agreed he then asked if he could speak to his wife. Berrett once again agreed. After a short conversation he offered to make a statement. In it, he implicated Browne in the murder of PC Gutteridge.

The ballistics expert Robert Churchill examined the weapons recovered from Browne and was able to prove, by the use of the comparison microscope, that the empty cartridge case retrieved from the vehicle had been fired from the Webley revolver found in Browne's possession when he was arrested. Browne's only possible defence against this evidence was that he had obtained the gun from Kennedy after the murder had occurred.

The trial opened at the Old Bailey on 23 April 1928 before Mr Justice Avery. Browne maintained his innocence of any involvement in the crime, claiming he was at home in bed that night. Evidence was heard from over forty prosecution witnesses, including four ballistic experts. The trial was notable for the forensic evidence given by Robert Churchill; through the use of photographs, Churchill proved to the court that the markings on the cartridge case matched those on the revolver, thus proving that the gun found in Browne's car was the murder weapon. Both men were eventually found guilty of murder and sentenced to death. They were executed on 31 May 1928.

George Gutteridge is buried in Warley Cemetery. The inscription on his headstone reads:

> In proud memory of George William Gutteridge, Police Constable, Essex Constabulary, who met his death in the performance of his duty on September 27th 1927.

The bullets and Webley revolver used to kill George Gutteridge are in the Essex Police Museum, while other exhibits relating to Browne and Kennedy are in the Black Museum at Scotland Yard. The road where Gutteridge was murdered has been altered since 1927, and a short stretch of it, now containing a memorial stone, has been renamed Gutteridge Lane in his memory.

Gunpowder was probably discovered in China in the ninth century, but it is not until the middle of the tenth century that there is the first visual evidence of a weapon using gunpowder. A picture that includes 'fire-lances' was painted on to a tenth-century banner from Dunhuang in Western China (see Plate 4). It depicts demons attacking with these fire-lances. In essence a fire-lance was a gunpowder-filled tube attached to the end of a long pole, used as a sort of crude flame-thrower. The Chinese also discovered that shrapnel rammed tightly down the barrel would fly out along with the flames and maximise the weapon's killing potential. An account of the siege of De'an in 1132 records that Song defenders used fire-lances very successfully against the Jurchen attackers.

There are a few theories as to how gunpowder came to Europe. One is that it made its way from China along the Silk Road, another that it came with the Mongols during their thirteenth-century invasion. The first recorded account of firearms being used in Russia states that during the defence of Moscow from Tokhtamysh's Golden Horde in 1382, the defenders used firearms called *tiufiaks*. During the fourteenth century in Europe, smaller and more portable hand-held cannons were developed. Such was the speed of development that by the late fifteenth century the Ottoman Empire was equipping its infantry with their own firearms.

Outside of warfare, however, guns tended only to be owned by the rich, mainly for hunting – after all, they were very expensive. It took several more centuries before they became the preferred weapon of criminals. The numerous wars in the seventeenth century, where firearms replaced more traditional weapons such as the longbow and crossbow, taught numerous less well-off people how to use guns, as well as flooding Europe with an accessible supply of them. This in turn led to a dramatic increase in crimes involving guns, especially the now much-romanticised highway robbery.

However, early guns were by no means without their operating problems: they were slow to load and only fired one shot, after which the whole laborious loading process had to begin again. A traditional longbowman could have fired off half a dozen arrows and an experienced crossbowman considerably more bolts in the time that it took to reload a firearm. There is a famous account written about an encounter between a highwayman and a young tailor during the eighteenth century. The tailor was crossing Hounslow Heath – a dangerous place at that time, infamous for highwaymen. Sure enough, he was attacked by an armed highwayman who demanded his purse. The tailor handed it over and then, thinking quickly, asked the highwayman to put a shot through his hat, so that it looked as though he had put up some sort of a struggle and would not be thought of as a coward. The highwayman obligingly did as requested. His single shot fired, he was, in essence, unarmed. The clerk now drew his own pistol and pointed it at the highwayman. He not only got his purse back, but also those of the other people the highwayman had already robbed.

When looking at forensic ballistics, it is useful to understand the principles that lie behind the operations of a firearm. From slow and unreliable beginnings, they soon developed into the efficient modern killing machines available today. Matchlocks were the first and simplest firing mechanisms developed for small arms. They were also slow to use. Gunpowder had to be poured down the barrel; this was followed by a lead ball, and then by some sort of paper wadding to keep the ball in place. All these then had to be rammed home by the use of a ramrod. The weapon was then primed by pouring powder into a pan at the back of the barrel. In the middle of the pan was a small 'touch hole', which connected to the powder in the barrel. The powder in the pan was then ignited by touching it with a lit match or piece of saltpetre-soaked string. A flame from the exterior gunpowder then raced inside the weapon through the touch hole to ignite the gunpowder in the barrel. This burned extremely rapidly, creating a large volume of hot gas that blew the ball out along the barrel, along with clouds of smoke. It was not uncommon for there to be so much smoke on the battlefield that it was impossible to see your enemy.

The successor to the matchlock mechanism was the wheel-lock, which was first seen in Europe in around 1500. Its invention is sometimes attributed to Leonardo da Vinci because of some drawings he made of such a mechanism in about the mid-1490s. Despite its many faults, the wheel-lock was a significant improvement on the matchlock in terms of both convenience and safety, since it eliminated the need to keep a smouldering match in proximity to loose gunpowder, which is never a good idea. It operated using a small wheel, much like that on cigarette lighters today, which was wound up with a key before use and which, when the trigger was pulled, spun against a flint, creating a shower of sparks that ignited the powder. Although the wheel-lock was a useful innovation, it was not widely adopted due to the high cost of the clockwork winding mechanism.

But it was the flintlock mechanism that really revolutionised small arms. The French artist and inventor Marin le Bourgeoys (*c.* 1550–1634) is credited with fixing the first flintlock mechanism to a firearm he had prepared for King Louis XIII in 1610. In this design, a sharpened piece of flint was clamped in the jaws of a 'cock' (a spring-loaded arm, so called because of its resemblance in shape to a rooster). When this was released by the trigger, it struck a piece of steel called the 'frizzen' to create the necessary sparks. The cock had to be manually reset after each firing, and the flint had to be replaced periodically due to wear. The flintlock was widely used during the eighteenth and nineteenth centuries in both muskets and pistols (see Plate 5).

Flintlocks were considerably more effective than what had

gone before, but still had a number of drawbacks, such as their unreliability in wet weather due to the flash pan of gunpowder being exposed. In approximately 1820, an invention by Reverend John Forsyth in Aberdeenshire addressed this issue. The percussion cap was a small cylinder, usually made of copper or brass, which contained fulminate of mercury, a shock-sensitive chemical compound that explodes when struck. The percussion cap was inserted into a hole at the back of the barrel of a gun. It was then struck by a spring-loaded hammer much like the cock of a flintlock, causing the fulminate of mercury to explode. The force of this explosion would travel down into the main barrel and ignite the gunpowder that had been loaded there, which in turn would propel the shot out of the barrel.

The appearance of the percussion cap was quickly followed by that of the cartridge, during the second half of the nineteenth century. Ammunition had previously always been delivered as separate bullets and powder. The cartridge combined a percussion cap, powder and a bullet in one completely weatherproof package. The main technical advantage of the brass cartridge case was the effective and reliable sealing of high-pressure gases at the breech (the part of a firearm behind the barrel), since the gas pressure forced the cartridge case to expand outwards, pressing it firmly against the inside of the chamber. This in turn prevented the leakage of hot gas, which could injure the shooter. The use of cartridges also opened the way for modern repeating arms, by uniting the bullet, gunpowder and primer into one assembly.

However, in spite of all these developments, accuracy was still an issue. The origins of rifling – whereby spiralling grooves are added to the inside of a gun barrel in order to impart a stabilising spin to the projectile – are difficult to trace, but some of the earliest practical experiments seem to have occurred in Europe during the fifteenth century. Early muskets produced large quantities of soot, which had to be cleaned from the barrel of the musket frequently, through either the action of repeated scrubbing, or a deliberate attempt to create 'soot grooves'. This might have led to a perceived increase in accuracy. True rifling, where the grooves are part of the inside of the barrel itself, dates from the mid-fifteenth century. This feature of firearms would become extremely significant for forensic ballistics, as we shall see later.

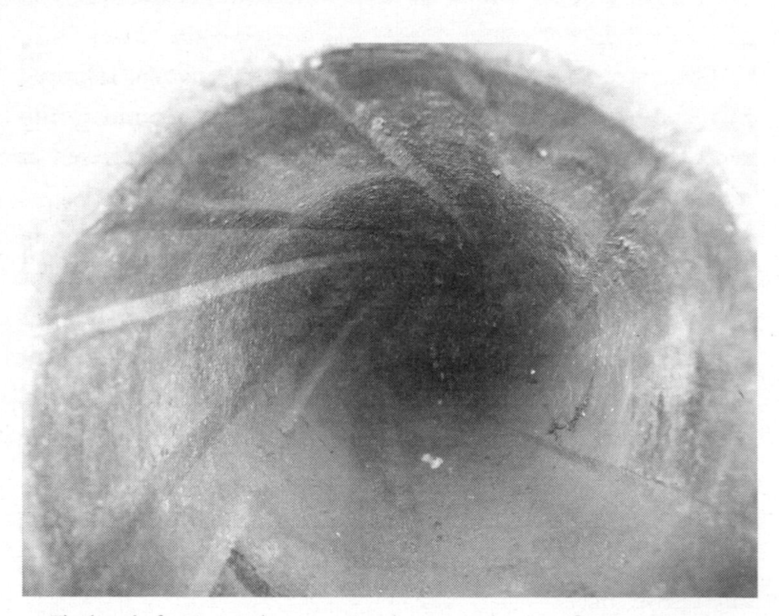

The barrel of a nineteenth-century French cannon, showing rifling in its interior. Rifling would later play a crucial role in allowing ballistics experts to match bullets with the guns that fired them.

In 1794 in Prescot, Lancashire, a man by the name of Edward Culshaw was shot through the head by a burglar. This murder was to make history as being the first ever to be solved by forensic ballistics. The local surgeon performed a postmortem on Culshaw. When he did so, he discovered not only the ball that had been fired from the pistol, but the remains of a small quantity of paper that had been used as wadding (this being the time of older, muzzle-loading weapons), which had been fired from the gun at the same time. When examined, it proved to be a small section from a song sheet. After some enquiries, an informant put forward eighteen-year-old John Toms as a possible suspect. Toms was arrested and searched, whereupon a ripped song sheet was discovered in his pocket. The wad of paper removed from Culshaw's head matched the damaged remains of the paper fired from the gun perfectly. Toms was tried, found guilty and sentenced to death. He was hanged for murder at Lancaster Assizes on 23 March 1794.

A similar case occurred almost a century later in France in 1891. Charles Guesner was a happy man, who had married the love of his life and settled down with her to begin a family. However, his happiness was to be short-lived. A few months later, a man broke into their bedroom while they were sleeping and, without warning, brutally shot Charles Guesner in the head. He died almost at once. The sound of the shot awoke his wife who, on seeing the bloody mess that had once been her husband's face, immediately passed out. The killer fled the bedroom, vanishing into the night.

The murder seemed as puzzling as it was brutal – Guesner had been a likeable and popular man. Nothing had been stolen, but for lack of a better explanation it was concluded that the crime had been committed by a burglar who had panicked after shooting Guesner and fled without taking anything. The only clue that the police recovered was the wadding that had been used in the gun. When they examined it, they discovered it was a page from the Lorraine almanac. Further investigation then revealed a suspect, a former suitor of Madame Guesner's by the name of Bivert. It was further discovered that Bivert had been very unhappy and jealous that she had married Guesner. The police searched his home and soon discovered the almanac, missing the page that had been used in the shooting. Bivert was tried and found guilty, and would normally have faced the guillotine. However, French law recognises a *crime passionnel* – an offence carried out in the heat of strong emotion – and, given his previous relationship with Madame Guesner, he was instead sentenced to twenty years with hard labour.

French chemist Roussin was to have the distinction of being one of the first scientists to solve a crime by using chemical analysis of a bullet. He did so in 1869. The victim of the shooting under investigation was the *curé* of Brétigny, who was shot through the head by an unknown assailant. Suspicion fell on a man named Cadet, a local watchmaker. Cadet was already known to the police and was no friend of the *curé*'s. The bullet had been recovered from the *curé*'s head, but had shattered on impact, breaking into numerous pieces. This meant that when bullets were discovered in Cadet's room, investigators had nothing against which to compare them. Nor did they have anything to match against the calibre (approximate inside width of the barrel) of the two pistols discovered in the room. This initially seemed like a major setback, until Roussin had the idea of chemical analysis. He established the exact weight of the bullets and their melting point, and was then able to work out their exact composition in terms of quantities of tin and lead. This done, he carried out the same analysis with the bullets discovered in Cadet's room. They matched. Cadet was tried and found guilty of the murder.

As with other branches of forensics, the science of ballistics has vindicated the innocent as well as convicted the guilty. In August 1876, a British police constable by the name of Nicholas Cock was shot and murdered in Whalley Range, Manchester. Two brothers, John and William Habron, were arrested and tried for the constable's murder. One of them, John, was acquitted, while the other, William, was convicted and sentenced to death (though this sentence was later commuted to penal servitude for life).

But this was not the end of the case. The day after the Habron trial, a man called Arthur Dyson was murdered in Ecclesall, a suburb of Sheffield. The killer was seen by the man's wife, who recognised him as Charles Peace, an infamous criminal with whom she had supposedly been having an affair. The motive for the murder was jealousy. Peace (1832–1879) was one of the best-known criminals of his day, and was even mentioned by Conan Doyle in the Sherlock Holmes story 'The Adventure of the Illustrious Client', as well as in Mark Twain's 'Captain Stormfield's Visit to Heaven'.

Peace evaded capture for two years after Dyson's death, but was finally arrested after shooting and wounding a police constable by the name of Robinson while attempting to burgle a house in Blackheath. He was tried and found guilty of Dyson's murder and was sentenced to death. Then, on his way to the gallows, Peace also confessed to the murder of Constable Cock and further said that he had acted alone. The police were not convinced, feeling that with nothing to lose he was just trying to get a fellow criminal off the hook. However, they did decide to compare the bullets taken from Arthur Dyson, PC Robinson and PC Cock. It was soon confirmed that they had all been fired from one gun, and that gun belonged to Charles Peace. William Habron was released from prison and compensated for his wrongful conviction and the three years of liberty he had been denied.

Across the Atlantic, America had also played a significant part in the development of ballistics. In 1830, a sixteen-year-old boy called Samuel Colt ran away from his home in Hartford, Connecticut. Like thousands of other boys before him, he went to sea, planning to see a bit of the world. It was while on his travels that he carved a wooden hand gun; it featured a revolving chamber that rotated through the action of cocking the hammer.

He returned to America and a few years later, aged only twenty-one, was producing working models of his gun from his factory in Paterson, New Jersey. Ingenious as they were, however, these prototypes were both complicated – having no fewer than twenty-four cogs, ratchets and springs – and, at $5 each, expensive. It was largely because of these issues that Colt's factory went bankrupt in 1842. However, by then his guns had already proved themselves. A year before the factory stopped trading, Texas ranger Jack Hayes was ambushed and cornered by a Comanche war party at a place called Enchanted Rock. The war party assumed that once Hayes' rifle had been fired it would take him a while to reload it and thought they could attack and scalp him during this time. They advanced carefully until Hayes finally fired his rifle, as they had expected him to. The Comanches immediately charged his position. However, unbeknownst to them, Hayes was also carrying a Colt revolver. He pulled it from its holster and killed them all.

When he heard of this new gun, Texas ranger Captain Hamilton Walker realised its potential and decided to lend Colt a hand. Colt had moved from Paterson to New York and Walker travelled there to meet him. Walker proposed that the two of them should go into business together to produce a new gun, based on the five-shot Colt Paterson revolver. Crucially, Walker offered to finance the operation. Colt accepted this offer of help and the partners built a new factory. By 1847 they were producing a newly enhanced revolver which featured a sixth chamber and which had just five moving parts instead of the original twenty-four. The United States Mounted Rifle companies were provided with the new revolvers, which proved to be extremely effective. The gun became known as the equaliser because of a popular poem written about it: 'Be not afraid of any man, no matter what his size. When danger threatens call on me and I will equalize.' It also made Colt and Walker millionaires.

With the Colt factory now turning out thousands of guns, they became cheap and easy to get hold of. It is no surprise, then, that they rapidly became the preferred weapon of bank and train robbers such as the infamous James gang, while gunslingers like John Wesley Hardin and Billy the Kid made their reputations as 'fast draws' with the weapon. Men like brothers Wyatt, Virgil and Morgan Earp earned equally keen reputations as lawmen. While there were undoubtedly downsides to this abundance of weapons, there was something about the design of the revolvers that would prove to be useful from a forensic point of view.

Revolvers included rifling as part of their design – the bullets were given spin by grooves that ran along the inside of the barrel. This meant that the outer case of the bullets picked up the shapes of the grooves, as well as any other imperfections the barrel might have. Alexandre Lacassagne (1843–1924), professor of forensic medicine at the University of Lyons, realised that this meant that each gun left a unique ballistic 'fingerprint' on every bullet it fired. In 1899, Lacassagne became the first known person to compare the markings on a bullet taken from the body of a murder victim to the rifling found in the handguns belonging to several suspects. During his examination of the bullet, he had discovered that it was marked with seven longitudinal grooves, created by its passage through the barrel. Only one of the weapons he examined had seven matching grooves inside the barrel. Fortunately, it was the gun that belonged to the suspect. As a result of this evidence, the killer was arrested and eventually convicted of murder.

But it wasn't until 1915 in Orleans County, New York, that ballistic evidence began to be taken seriously in the United States. It was the infamous case of Charlie Stielow and Nelson Green that proved just how vital it could be in establishing a person's guilt or innocence.

During the early morning of 22 March that year, a

good-natured but illiterate farmer by the name of Charlie Stielow discovered the body of a dead woman on his front doorstep, still dressed in her nightgown. When Stielow looked more closely, he discovered it was Mrs Margaret Wolcott, the housekeeper of a man called Charles Phelps. Phelps was Stielow's boss and the owner of the farm he worked on. Margaret had been shot in the chest. It happened that it had snowed overnight and Margaret Wolcott's footprints were still fresh in the snow. Stielow decided to retrace her steps, whereupon he discovered that she had come from Phelps' home. He found the kitchen door wide open and, on entering, found ninety-three-year-old Phelps lying on the floor, having been shot three times. Incredibly, he was still alive, though only just. The local police were immediately called and quickly arrived on the scene. However, not being used to dealing with crimes of this magnitude, they almost certainly did more harm than good, tramping through the crime scene, destroying and disturbing evidence. In the end, the Orleans County authorities took the unusual step of employing George W. Newton, a private detective from Buffalo, New York, to investigate the killing. After an investigation lasting only ten days, Newton suggested that the culprit was a man by the name of Nelson Green, who was Stielow's brother-in-law and who lived with Phelps and his wife. He was arrested and before long confessed to the murders, implicating Stielow at the same time. Stielow was then arrested and he too quickly confessed. Interrogation techniques at the time were far more physical than they are today and beating a confession out of someone was pretty common.

In their written confessions – which were very much

alike – the two said that they had knocked on the kitchen door and shot Phelps when he answered it before making for the bedroom, where they believed Phelps kept his money. While they were doing this, Mrs Wolcott ran from her room and dashed out through the kitchen door, slamming it shut behind her. At this point either Green or Stielow (it is impossible to say which since they both confessed to doing all the shooting) apparently opened fire at the housekeeper through the glass panel in the kitchen door, cutting her down before returning to the bedroom to search for the money. Each man's statement then said that he had found $200, which he had given to the other to keep. Both agreed that when they returned home they heard Mrs Wolcott calling for help but ignored her pleas and left her to die. Newton also discovered that both men had lied about owning guns, and that Stielow had given both a .22 calibre revolver together with a .22 rifle to a relative to hide. The victims had been killed with a .22.

Stielow was tried first. He retracted his confession, saying that he had been forced to sign it, but in spite of this it was admitted into evidence. To tie everything up, all that was needed was to link the bullets removed from the bodies of the victims to one or both of Stielow's guns. The prosecution brought in the self-styled ballistics expert Dr Albert Hamilton from Auburn, New York. He was, in fact, a fraud – he styled himself a doctor but was in reality a hawker of patent medicines with no qualifications at all in any field of science or medicine. Since nobody had ever challenged him, he was able to become an expert witness in everything from toxicology and bloodstains to handwriting, and also conducted autopsies. For the right money he would go along with anything the police told him to say, and unfortunately his lack of any real knowledge was not discovered for a long time.

In his evidence 'Dr' Hamilton testified that he had microscopically examined Stielow's revolver and discovered nine defects at the end of the muzzle that matched nine scratches found on each of the four bullets taken from the victims. When cross-examined and asked why these scratches didn't show up on the enlarged photographs of the bullets, he simply replied that due to some error the marks were on the opposite side of the bullets to those photographed. For some extraordinary reason nobody, not even the defence, asked him to provide new photographs where the scratches were visible; they just accepted him at his word. The defence then asked why it was that only the defects at the very end of the barrel would mark the bullets. Hamilton was a seasoned conman and could sound very convincing when he needed to. He blinded them with pseudo-science. This was his reply: 'The cylinder fitted so tightly against the rear of the barrel that there was no leakage of gas at the breech. The full force of the gas following the bullet out at the muzzle, the lead expands as it leaves the muzzle, fills in any depressions existing at the outer edge of the bore and receives scratches from the elevations existing between said depressions.'

Perhaps this explanation seemed detailed enough to be plausible, since no one challenged it. The jury found Stielow guilty of murder in the first degree and he was sentenced to death in the electric chair. Green, wishing to avoid the same fate, pleaded guilty and was sentenced to life imprisonment. In February 1916, an appeals court upheld the convictions, stating that 'from an examination of the record, it is inconceivable that the jury could have rendered any other verdict'.

While awaiting execution at the infamous Sing Sing Prison, Stielow managed to convince the deputy warden, Spencer Miller Jr, that he was innocent of the crime. Miller passed on his concerns to Louis Seibold, a reporter with the *New York World*. He in turn hired a Buffalo detective called Thomas O'Grady to re-examine the case.

O'Grady discovered that both the defendants were illiterate and would therefore have been incapable of writing out their own confessions. He also found it difficult to believe that they would have used some of the sophisticated phrases found in their statements. He then discovered that both Newton and Hamilton were working on a contingency basis – that is, they would not have been paid unless both Stielow and Green were found guilty.

O'Grady's investigation continued, but meanwhile the second and third appeals for a new trial were denied. Time was now running out for Stielow. Fortunately, by now the state governor had taken an interest in the case and on 4 December 1916 he commuted Stielow's sentence to life imprisonment. He also appointed a Syracuse attorney named George Bond to re-examine the case. Bond promptly employed yet another investigator, named Charles Waite, to do the 'leg work'.

Bond and Waite quickly established that the confessions made by the two men did not correspond convincingly with the facts of the case. The statements that Stielow and Green had signed said that Margaret Wolcott ran past them, but if that was so, then she must have recognised them, as she knew them both well. Why, then, would she have run to Stielow's house to find help, knowing that he was one of the men who had attacked her? Both men had also said in their confession that Margaret Wolcott was alive when they emerged from Phelps' bedroom and ran back to Stielow's house, yet given that she had been shot in the heart, this seemed more than a little unlikely. And if these circumstances were not bizarre enough, it was also noticed that the angle at which the bullet had entered her body was geometrically impossible given the immediate geography, as anyone visiting the scene would have noticed at once.

Stielow's .22 revolver was then examined by ballistic experts from the New York Detective Bureau. It was their expert opinion that the gun had not been fired for at least three or four years. They wrapped a sheet of paper around the gun and fired a single round. The paper burst into flames, ignited by the hot gases discharged – an obvious contradiction of Hamilton's statement that there was no leakage of gas at the breech. Next the gun was discharged into a cotton-filled box and the bullets recovered and taken to Dr Max Poser, an expert in microscopic examination at the Bausch and Lomb headquarters in Rochester. Not only was Poser unable to find the microscopic scratches that Hamilton had sworn in court were there, he also discovered that the bullets had been fired from a gun with a manufacturing flaw. One of the five grooves that were supposed to line the inside of the barrel of that particular model was missing. Stielow's gun had no such defect, and therefore could not have fired the bullets that killed Phelps or Margaret Wolcott. When this evidence was presented, both Stielow and Green were pardoned. They were released on 9 May 1918.

While forensic ballistics had certainly helped to save the life of a man who had been falsely accused, the conclusion of this case is perhaps not as satisfying as we would like. A man called King came to be suspected of the crime instead, but was never prosecuted; neither were Hamilton and Newton, who had lied under oath and thereby almost caused the death of an innocent man. Stielow and Green were never compensated. Alas, true stories never have quite the ending that we would like them to.

Still, as a result of this case, the American authorities began to see that the ability to accurately match the bullets taken from a crime scene with a particular gun had extremely useful practical applications. Charles Waite, whose involvement in the Stielow and Green case we have just mentioned, began to assemble data on all manufactured guns – their bore diameters, the pitch and direction of their rifling and anything else that might help match a bullet with the gun that fired it. A survey of only American-made guns soon proved to be insufficient. At the end of the First World War, America was flooded with cheap, mass-produced foreign pistols, mostly from Europe. In order to broaden his database to accommodate the circulation of these weapons, Waite therefore travelled to Europe and spent the bulk of his time there for several years.

During the 1920s, while Waite was compiling his database, ballistic expert Calvin Goddard and chemist Philip Gravelle were busy perfecting the comparison microscope. This is a binocular device where each eyepiece views a different area through a separate microscope. A simple early model had already been used to compare such things as grains and ground pigments. Goddard and Gravelle modified it so that bullets or shells could be compared side by side. It was a revolution in the science of ballistics. In 1925, Goddard and Gravelle teamed up with Waite to establish the now legendary Bureau of Forensic Ballistics in New York. From here they offered their services to police forces throughout the country, specialising not only in ballistics but also in fingerprinting, blood-typing and trace evidence analysis – in just about any forensic technique, in fact.

Probably the bureau's most famous case took place on St Valentine's Day 1929. On that day, at a garage situated at 2122 North Clark Street, Chicago, two men dressed as police officers lined six members of the George 'Bugs' Moran gang up against a wall. Two other men joined them, both wearing trench coats and carrying Thompson sub-machine guns. They then proceeded to fire seventy rounds into the line of men, killing some instantly and seriously wounding others (of those wounded, none survived very long).

One of the victims was a gangster named Frank Gusenberg. While he was still lying on the garage floor, a genuine police officer who had arrived at the scene told him he was dying and asked him to name the person who shot him. He replied: 'I'm not going to talk – nobody shot me.' He died with seventeen bullets inside him. The criminal code of silence had been maintained even under the most extreme circumstances.

Just over a year after these events – now infamous as the St Valentine's Day Massacre – two Thompson sub-machine guns were recovered from the home of a known hit man called Fred Burke, who had been arrested on suspicion of murdering a police officer in Michigan. Goddard compared the bullets from the massacre with test bullets fired from the recovered Thompsons. They matched; they had their murder weapons.

Frustratingly, despite strong evidence against Burke and his gang, for some unknown reason (probably corruption) he was never charged with the crime. The case remains, officially anyway, unsolved. Unofficially it is almost certain that Burke and his gang were responsible, and it is only because of the work of the Bureau of Forensic Ballistics that we are able to make that statement. Burke was later convicted for the murder of the police officer, and died in prison in 1940.

As firearms have been improved and refined, so ballistics experts have had to change their methods in order to keep up. There are certain characteristics that all similar weapons will have in common – things such as the calibre of the bullet, the number and size of the rifling grooves inside the barrel, and the position of the marks on the shell. These are known as 'class characteristics'. All similar weapons will have the same class characteristics; for example, the barrels of all .45-calibre Colt automatic pistols have six rifling grooves with a left-handed twist. The groove depth in the Colt .45 is .0035 inch, and the rate of twist is one full turn in 16 inches. Calibre is the measure of the diameter of the bore (the interior of the barrel) in hundredths of an inch – a .30-calibre gun has a bore of 30 hundredths of an inch. Unfortunately, this simple system of categorisation has slipped a little over time. Thus the .38-calibre Colt special has a bore of only .346, and the so-called .38–.40 has a bore of .401. To further complicate matters for the ballistics specialist, over the years many small-arms manufacturers have made speciality guns of unusual calibres. Still, despite these complications class characteristics can usually determine the model of gun from which a particular bullet was fired.

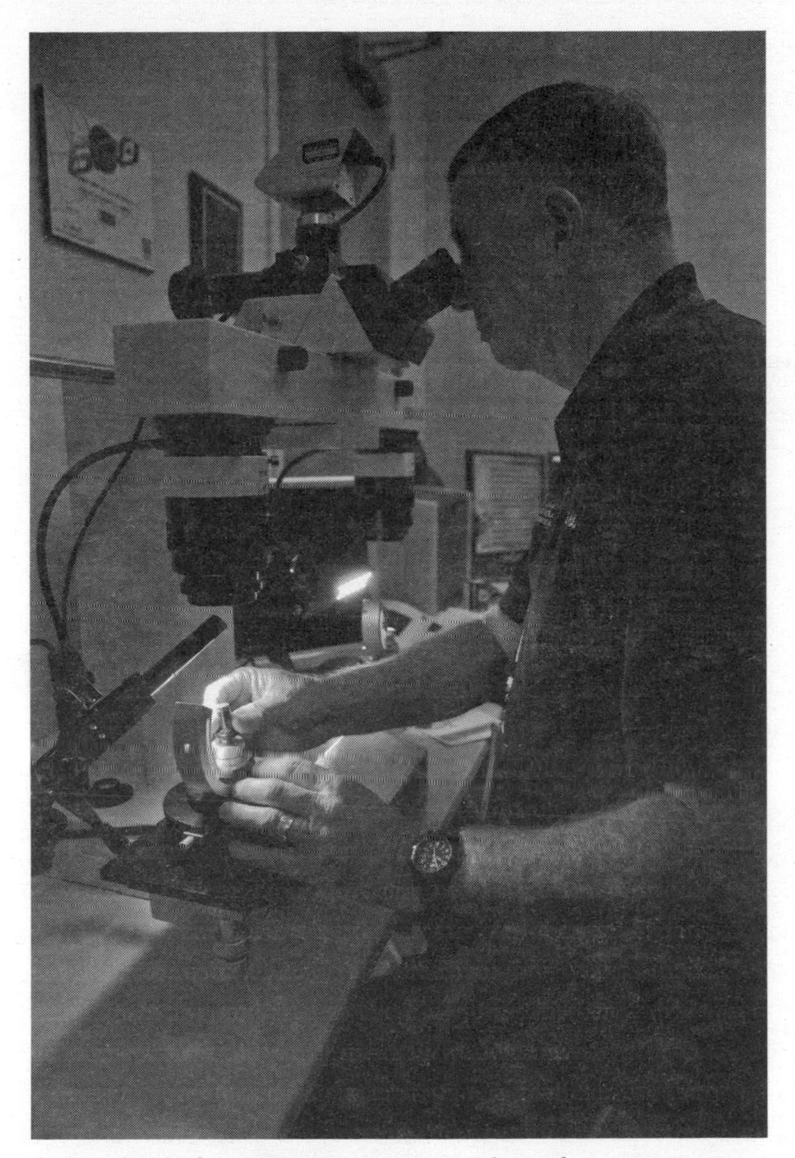

A modern-day forensic investigator mounts a cartridge case from a crime scene on a comparison microscope at the Santa Ana, CA, police department. Microscopes such as these were instrumental in allowing simultaneous comparison of cartridges.

The variety of firearms available means that determining the origin of a bullet requires an understanding of class characteristics as well as an extremely current knowledge of the range of weapons on the market.

Even more complicated details such as the rate of twist of the barrel can be determined using some reasonably straightforward calculations. First, you must measure the diameter of the bullet and the angle the groove makes relative to a straight line drawn from the point of the bullet to the back. The formula for determining the twist of the barrel from which a bullet was fired in inches is:

$$P = \pi \times D \div \tan a$$

P is pitch (meaning the twist), D is the diameter of the bullet, and *tan a* is the tangent of the angle of the groove. Suppose you're looking at a .45-calibre bullet with a diameter of .451 inch and you find the angle of the groove to be 5° 04'. Your scientific calculator tells you that the tangent of 5° 04' is .0885. You multiply π (3.14159) times the diameter and get 1.4168. You then divide that by .0885, which tells you that the twist is one turn in 16 inches.

But after an expert has managed to identify the type of gun that fired the shot, there is still one question remaining: which gun in particular was it? Fortunately, there are ways to establish this, related to the process of manufacturing a gun. To create the barrel, a bore is reamed out of a solid metal rod and the tool that does this job leaves behind myriad tiny scratches. A smoothing tool then reduces these scratches to microscopic proportions, but, crucially, they do not disappear altogether. Another tool then cuts the grooves in the barrel, and this process leaves behind its own pattern of tiny scratches. Additionally, each cut creates minute changes in the cutting tools themselves, meaning that the structure of each barrel is slightly different. What all this means is that every barrel is individual, and leaves a different pattern of striations on the bullets that pass through it, even if these are only visible on a microscopic level. By test-firing bullets from a weapon they have reason to suspect has been involved in a crime, and then comparing these with bullets recovered from the scene, an expert can establish whether the microscopic markings match, and therefore whether the gun is indeed the one they are looking for.

The importance of the unique marks that an individual weapon can leave upon the projectiles it fires is well illustrated by the case of Dr Angelo Zemenides. Zemenides was a Cypriot who lived in London and worked as a teacher, as well as being an interpreter at the Old Bailey for the police. As a result of this latter occupation he had received several death threats. Zemenides wanted to boost his income still further and so resolved to become a marriage broker. He accepted a £10 fee from one Theodosios Petrou – a fellow Cypriot who worked as a waiter in an upmarket restaurant in Piccadilly Circus – on the condition that he find Petrou a bride with a £200 dowry. Since after some time no bride was forthcoming, Petrou asked for his money back. Unfortunately Zemenides only had £5 left. This he handed over, explaining rather feebly, 'I spent the rest.' Petrou was unsurprisingly furious.

On the evening of 2 January 1933, Zemenides was at his lodgings in Hampstead and had settled in for the evening. Around 11.20 P.M. there was a knock on the front door of the lodgings. It was answered by another resident, a Mr Deby. The man at the door asked to speak to Dr Zemenides and Deby allowed him in. A few minutes later there was the sound of a struggle followed by several shots. The man Deby had admitted fled into the night. Zemenides was found inside his room. He had been shot dead.

Given the circumstances, Petrou was an obvious suspect. The police took him into custody. When they searched the cellar where Petrou lodged, they recovered a .32 self-loading Browning pistol with five cartridges still in the magazine. Two of these cartridges were standard self-loading pistol cartridges, rimless with nickel-jacketed bullets; the other three were .32 revolver cartridges with the rims filed off and with lead bullets. A search of the crime scene resulted in the recovery of two fired cartridge cases, one a rimless self-loading pistol cartridge, the other a .32 revolver cartridge with the rim filed down. The bullet removed from Zemenides' body was a nickel-jacketed pistol bullet. A second projectile recovered from the panelling at the scene of the murder was found to be a lead revolver bullet. It was assumed by the investigating team that the recovered bullets had been fired by the gun found at Petrou's lodgings.

The defence retained the services of Major Sir Gerald Burrard, who, like the famous Churchill, was one of the country's leading firearms experts. With the assistance of the handgun expert Dr R. K. Wilson, he began work on the cartridge cases. After lengthy detailed examination, the two ballistics experts finally managed to demonstrate that the gun found at Petrou's lodgings could not have been the murder weapon because the bullets removed from the body and those found at the scene did not match. The jury were convinced by this evidence and returned a 'not guilty' verdict. The real killer was never discovered and the crime remains unsolved. The circumstantial evidence against Petrou was strong and, had it not been for the ballistic evidence supplied by Burrard and Wilson, he might well have been hanged for a crime he did not commit.

The invention of guns not only revolutionised warfare but also made the job of the criminal a great deal simpler – a person can easily carry a concealed pistol, making it a convenient and horribly efficient way to kill. In the modern world more people are murdered by handguns than by any other single means. The analysis of bullets and weapons is therefore a necessary and often extremely significant forensic skill. However, as with so much forensic science, it is a game of cat and mouse – as soon as the scientists solve one problem the criminals will become aware of their progress and react accordingly. Ensuring that no shell cases are left lying around, making sure that the bullet shatters on impact and destroying the gun after use are all methods employed by criminals today in order to hide their tracks. It's an ongoing battle, but one that forensic scientists remain determined to win.

3

Blood

Who would have thought the old man to have had so much blood in him.

William Shakespeare, *Macbeth*,
Act V Scene I (*c.* 1606)

Addine G. Erskine commented in his book *The Principles and Practice of Blood Grouping* that: 'for at least as long as recorded history, man has been interested in and mystified by blood'. It's true that there can be few substances that carry such symbolic weight: blood sustains life, but as it is only ever seen when a body has been damaged, paradoxically we also associate it strongly with death. Since violent crime almost inevitably results in blood being spilled, it is perhaps unsurprising that the study of blood has been one of the most important aspects of forensics for a long time.

That said, it wasn't until the early part of the twentieth century that blood analysis began to play a really significant part in crime investigation (see Plate 6). Prior to that time the knowledge it was possible to gain from analysing blood was limited; there wasn't even a method for distinguishing human blood from animal blood until 1901. An incident in Scotland in 1721 serves to illustrate the problems that could arise as a result of this lack of knowledge.

William Shaw lived in Edinburgh. He had a daughter called Catherine, though it was widely known that he wasn't on the best of terms with her, largely because he disapproved of the man she had been seeing. One evening, neighbours living within the same tenement heard a violent argument going on in the rooms where the Shaws lived. It concluded with several groans and the sound of a door slamming, followed by silence.

Concerned, the neighbours decided to knock to check that all was well. When there was no reply they sent for the police, who forced open the locked door upon arrival. Inside they were greeted with a terrible scene: Catherine Shaw was lying in a pool of blood, a bloody knife at her side. She was still alive but unable to speak. However, when asked if her father was responsible for her present condition, she nodded her head. She died shortly afterwards without being able to give a fuller account of what had occurred.

A little later, Shaw returned to the flat. The police found bloodstains on his clothing and arrested him immediately. He was charged with the murder of his daughter soon afterwards. In his defence he said that Catherine had committed suicide out of despair at being unable to be with the man she loved (because of his refusal to allow them to marry). He admitted that they had argued, fiercely, but he said that he had not harmed her, having stormed out of the room in a rage. He further claimed that the blood discovered on his clothes was his own, saying that he had cut himself a few days before and that the bandage had come loose, letting blood drip onto his clothes. However, the jury was not impressed by this explanation and Shaw was found guilty and sentenced to death. He was hanged in November 1721, still protesting his innocence to the last.

Considering that both father and daughter were dead, one would expect that to be the end of the matter. However, the next tenant to move into the flat discovered a letter in a small opening near the chimney. When it was opened and read, it turned out to be a suicide note from Catherine. She said she had decided to kill herself because her father would not allow her to marry the man she loved, and concluded by saying that, as a result, he was the cause of her death. When the letter was examined and its authenticity established, the authorities realised they had hanged an innocent man. William Shaw's body was cut down from the chains in which it had been hung and he was given a Christian burial – the very least they could do.

Today it is unlikely that a miscarriage of justice such as this could occur: modern science enables us to establish a blood type, so it would be possible to verify that the blood on Shaw's clothes was indeed his own and so confirm the truth of his story, or at least that part of it. However, in 1721 such technology was still a long way off.

The first really significant advance in blood analysis was made in 1853 by the Polish physician Ludwig Teichmann. He developed a test to confirm the presence of blood which, though complex, was nonetheless effective. He discovered that if you dissolved a sample of dried blood in a mixture of saltwater and glacial acetic acid, then warmed the mixture, microscopic prismatic crystals would form; a substance he called hematin. A version of this test is still used today to identify whether dried stains found at a crime scene contain blood.

Hematin extracted from the blood of cattle; the crystalline structure which Teichmann observed is visible.

A decade later the German chemist Christian Friedrich Schönbein – who is also the discoverer of ozone and the inventor of the fuel cell – discovered that hydrogen peroxide will foam in the presence of blood. Even a tiny amount sets off this reaction. The downside was that the same thing happens with small amounts of semen, saliva, rust and some kinds of boot polish, since all these substances contain specific enzymes that cause the oxidation of hydrogen peroxide. Despite this drawback, Schönbein's test was still useful, as it provided a quick way of eliminating suspicious stains – if the hydrogen peroxide did not foam, then you at least knew that blood was not present, and if it did then you knew that further investigation was required.

Other scientists continued to work on the same problem, and by the end of the nineteenth century there was a wide array of tests to detect the presence of blood. However, there was still no test that differentiated between human and animal blood. In 1841 a French chemist named Barruel believed he had cracked this problem; he claimed that when heated with sulphuric acid, human blood gave off a specific 'sweat' odour that was unique to it. Some Paris courts were convinced by Barruel's claims and allowed information derived from his methods into evidence. Unfortunately his theory had no basis whatsoever in fact. In 1850, Ludwig Teichmann also thought he had found a solution to the puzzle when he developed a test based on the shape of blood crystals. Unfortunately, as well as being rather complicated, this test was fairly prone to error and was therefore of very limited practical use.

Extraordinarily it was Sir Christopher Wren (1632–1723), the architect of St Paul's Cathedral and many other magnificent buildings, who had paved the way for the mystery to be solved many years before. He had studied at Oxford University and was a respected scientist. In 1656 he invented the intravenous injection. The syringe he used was a quill with a sharp point, attached to a bladder. Instead of the device piercing the skin, an incision had to be made to make the vein available – as basic as it comes (and certainly not for the squeamish), but it worked.

The hypodermic syringe, as we recognise it, did not appear until 1853 (Dr Alexander Wood of the Royal College of Physicians of Edinburgh is generally credited with its development), but the existence of at least crude syringes meant that in 1814 Dr James Blundell (1791–1878) was able to start experimenting with blood transfusions. The first transfusions had already occurred in 1667 in Paris, performed by Jean-Baptiste Denys (1643–1704). Although Denys believed that transfusions should be performed using human blood, he decided to use animal blood, as he considered the risk to the blood donor was too high. His transfusions therefore had fatal results, which led to the practice being banned in France and England. It was Blundell who reintroduced it. Blundell was fascinated to discover that he could almost drain a dog's blood before reviving it with the blood from another dog. If, however, he used the blood from a creature of a different species, such as a sheep, the dog died. By 1818 Blundell was trying blood transfusions on human patients, but was unable to understand why some lived while others died after receiving the same treatment.

It was German physiologist Leonard Landois (1837–1902) who came up with at least a partial explanation for this. He observed that if red blood cells from one animal were mixed with the serum – the liquid base of blood in which the cells are suspended – of an animal of a different species, the red cells clumped together like lumps in porridge, and on occasion even burst. It was clear that if this reaction occurred in the human body, it would result in fatal illness.

However the breakthrough finally came when Karl Landsteiner (1868–1943), then an assistant professor at the Institute of Pathology and Anatomy in Vienna, published a paper in the *Vienna Clinical Weekly* entitled 'On Agglutination Phenomena of Normal Human Blood'. The paper described the results of experiments he had done using his own blood along with that of several colleagues. In these experiments he found that mixing blood serum from one human being with the blood serum from another would sometimes produce the same 'clumping' reaction or (to use the proper scientific term) agglutination. But the questioned remained – why?

He concluded that as some blood samples caused other blood samples to clump together, there must exist at least two 'antigens', which he named anti-A and anti-B. Blood serum contains specific antibodies that react with different types of antigen. In fact, Landsteiner eventually concluded that there were four blood types, which he named A, B, AB and O. The letters indicate the different types of antigen (which are a kind of protein) found on the surface of red blood cells of different types. Type A blood cells have the A antigen, type B have the B antigen, type AB have both and type O have no antigens. Blood serum contains specific antibodies that react with different types of antigen. As a result, type A serum (or rather the antibodies within it) will agglutinate type B blood and type B serum will agglutinate type A blood. Type AB will agglutinate both type A and type B. However as type O has no antigens, it can be mixed safely with any serum type. It is only safe to give a patient a transfusion of blood if it is of a type that does not agglutinate with their own.

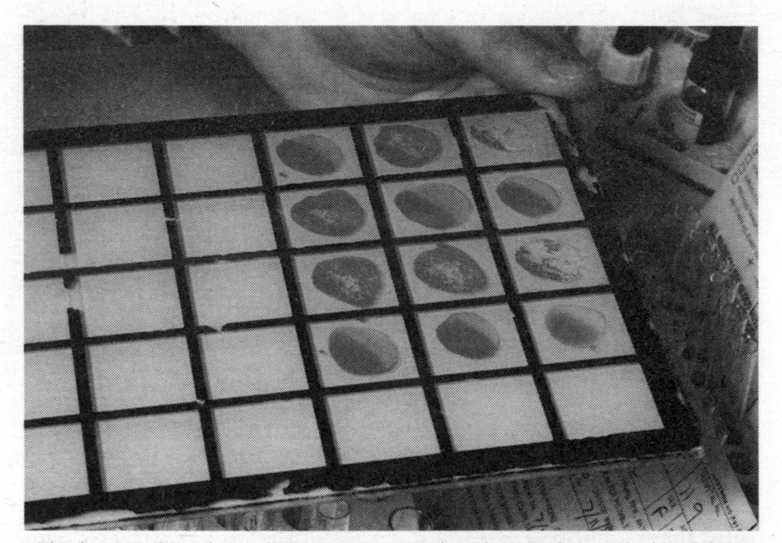

Thanks to Karl Landsteiner's discoveries, in modern-day transfusions the blood from the donor and that of the recipient are carefully compared to ensure that the blood types are compatible and will not agglutinate upon transfusion.

At the same time that Landsteiner was conducting his experiment, a young doctor by the name of Paul Uhlenhuth (1870–1957), at the Institute of Hygiene in Greifswald, discovered how to distinguish between animal and human blood. It started with a significant discovery on the part of German physiologist Emil von Behring (1854–1917). In 1890 he found that animals inoculated with diphtheria toxin formed defensive substances in their blood serum. In 1900, building on the work of Behring and others (such as the Belgian immunologist and microbiologist Jules Bordet, 1870–1961), Uhlenhuth discovered that if he injected protein from a chicken egg into a rabbit, and then mixed the blood serum from the rabbit with egg white, the egg proteins separated from the liquid to form a substance known as a precipitin. However, this only worked with chicken egg white – whites from any other bird would not precipitate chicken protein. Uhlenhuth then tried to create a serum using chicken blood instead of egg white, which again caused the protein to be precipitated out. He had therefore created a serum that precipitated the protein of only one animal. Using the same technique he began creating serum tests for every possible animal. These discoveries enabled him to differentiate not only between animal blood types, but also between human and animal blood.

Uhlenhuth then went on to develop safeguards for the test. When faulty results were obtained through the use of serum that had been created in another lab, he standardised the serums by insisting that the only official sources should be his own institute and the Robert Koch Institute in Berlin. He also strongly recommended that before any test of an unknown substance, a control test should be conducted against a known sample. When testing a particular bloodstain, it might also be the case that misleading results could be obtained as a result of the underlying material on which the stain lay. Uhlenhuth therefore suggested that any such material should be tested separately first, in order to eliminate the possibility of a false positive. With these safeguards in place, the Uhlenhuth precipitin test worked infallibly every time. Its practical value was first demonstrated in 1901.

During the early afternoon of 9 September 1898, two little girls went missing from the village of Lechtingen, near Osnabrück, Germany. Their disappearance naturally caused a great deal of anxiety and so a search was mounted by the whole village. Towards evening the body of one of the girls, seven-year-old Hannelore Heidemann, was discovered in some nearby woods. It was a gruesome sight: her body had been dismembered and parts of it scattered around. An hour or so later the remains of her friend, eight-year-old Else Langmeier, were found hidden under some bushes. She had also been mutilated and dismembered.

A local carpenter by the name of Ludwig Tessnow quickly fell under suspicion, as he had been seen entering the village from the direction of the woods and appeared to have bloodstains on his clothing. He was arrested and questioned by the police, but claimed that the brown marks on his clothes were simply stains from wood dye, which he used in his work. With no way of proving otherwise, the police had to accept this explanation and Tessnow was eventually released. He remained in the district for some time, taking work where he could find it, before wandering further afield in 1899. He finally settled in the small village of Göhren on the Baltic Island of Rügen.

On Sunday 1 July 1901, two young brothers from the village,

Peter and Hermann Stubbe – six and eight years old, respectively – were found murdered and disembowelled. Their heads had been removed, their skulls crushed with a rock and their limbs severed. Hermann's heart had also been removed and was never discovered.

Once again there was cause to suspect Tessnow. One witness recalled seeing him talking to the two boys earlier that day, while another remembered him returning to the village in his Sunday best with dark stains visible on both the jacket and trousers. When the local authorities interviewed him he denied any involvement, but a search of his home unearthed recently laundered clothing that bore suspicious stains. As before, he claimed that they were from wood dye and, as before, he was released. However, one of the magistrates remembered Tessnow's name being mentioned in connection with the Lechtingen murders a few years earlier. A local farmer had also seen a man whose description matched Tessnow run from his field, leaving behind him seven slaughtered sheep. The sheep hadn't just been killed; their legs had been cut or torn off and tossed about the field. When Tessnow was brought in for a line-up, the farmer had no trouble picking him out as the man he had seen.

Nevertheless, the police needed real evidence in order to tie Tessnow to the murders. Uhlenhuth had developed his test to distinguish animal blood from human only four months earlier. When the authorities heard about this, they contacted him and asked him to test Tessnow's clothing and the bloodstained stone used to crush the children's skulls. Uhlenhuth had been preparing for such a test, and applied his method to more than one hundred spots. He then announced the results: while he did find wood dye, he also detected traces of both sheep and human blood, all quite distinct from one another. Tessnow was tried, convicted and executed on the strength of this evidence.

It was almost a decade later that blood analysis played an important part in solving a murder in England for the first time.

In July 1910, seventy-year-old widow Isabella Wilson was found dead in the back room of her second-hand clothes shop on the High Street, Slough. A cushion had been tied tightly over her face with a scarf so that she suffocated. Injuries were also discovered on the side of her head, indicating that she had been struck several times with some kind of blunt instrument. The motive for the killing seemed clear – it was generally known that Mrs Wilson kept a purse in the pocket underneath her apron and rumour had it that she sometimes carried as much as twenty gold sovereigns in it. The purse was discovered next to her body, empty. On the table nearby, the police found a piece of brown paper with circular marks on it. Mrs Wilson was known to keep her sovereigns wrapped in brown paper – clearly the killer had left the paper behind.

Their enquiries soon pointed the police to a man called William Broome, a twenty-five-year-old unemployed motor mechanic. He had once been a neighbour of Mrs Wilson's, though had moved with his family some time previously. On the day Mrs Wilson died, however, he had been seen in Slough by several people. Broome was tracked down in Harlesden a few days after the murder. When he was taken to a local police station and questioned, he claimed to have been in London all day, contrary to the statements given by a number of witnesses placing him in the vicinity of the crime. When asked if he had any money on him, he produced only a few shillings but then, quite extraordinarily, explained that he had twenty gold sovereigns back in his room in Albany Street, Regents Park.

Broome also had two scratch marks on his face. He claimed he had received them during a fight with a local bookmaker over a bet on a race he had won. This was also where he had got the twenty gold sovereigns, he said. However, Mrs Wilson had long nails, one of which was broken. It had been assumed that this had occurred when scratching her attacker in an attempt to defend herself.

Broome was committed for trial at Aylesbury assizes on 22 October 1910. He maintained his story, continuing to insist that he had been in London at the time of Mrs Wilson's murder. The prosecution was already able to produce several witnesses who knew Broome well, and who were adamant that they had seen him in Slough, but what really caused a stir was Dr William Willcox taking the stand – he was already well known for giving evidence in the infamous Dr Crippen case. The investigating team had asked Willcox to examine several fingernails snipped from the late Mrs Wilson's fingers at the scene, and also to look at a pair of Broome's boots and some of his clothing.

Willcox noted that there was skin attached to one of Mrs Wilson's fingernails and that it had blood on it. When he turned his attention to the boots, he found that, although Broome had carefully cleaned and polished them, he had failed to notice stains on the instep. Willcox examined them and said that they were blood of mammalian origin. Asked if this meant the blood could have come from a human, he replied firmly that it could, though he did not attempt to establish himself whether the blood was human or animal, or what group it was. Willcox probably assumed that with the weight of evidence against Broome, he didn't have to spend much time investigating such details.

He did, however, go on to examine the brown paper found at the murder scene under a microscope. When he did so, he discovered small specks of gold. Willcox did not merely say this proved that the deceased did indeed wrap her coins in the paper, he made the rather outrageous claim that he was able to tell that the paper had contained twenty gold sovereigns, the exact amount found in Broome's room. Willcox was an enormous presence in court and juries would watch him with rapt attention, as though witnessing a magician perform a marvellous trick. His evidence certainly sealed Broome's fate – once the judge had summed up, it took the jury only thirteen minutes to find him guilty as charged. He was sentenced to death.

The next notable advance in serology took place in Italy in 1915, when Dr Leone Lattes, an Italian lecturer in forensic medicine in Turin, became determined to prove that you could establish blood groups even weeks after stains had been discovered, long after the blood had dried. His opportunity came in a rather bizarre set of circumstances, when he was asked to settle an everyday domestic dispute.

A construction worker named Ranzo Girardi returned home from a trip to another town with what appeared to be bloodstains on his shirt. His wife saw them and, in some kind of strange paranoia, accused him of adultery during his trip. He was unsure where the blood had come from, but vehemently denied her accusations. However, she refused to believe him and began to make his life a misery. After three months, Girardi was growing desperate and so consulted the Institute of Forensic Medicine for help. Lattes, who worked there, agreed to test the stains and to try to establish a match, even though they were far from fresh. Girardi thought the stains might very well be his own blood, though it was also possible they were his wife's or possibly even beef blood from the butcher.

Lattes soon determined that the stains were human, eliminating the latter possibility. He then established that Girardi's blood was type A and that his wife's was type O. He also took the blood of a friend of Girardi's wife, who had been staying at the house at the time and who had been menstruating, his reasoning being that she might have inadvertently transferred some of her own blood onto the shirt. This friend was also found to be blood group A. Despite the fact that the outcome of this marital dispute was of no real significance, Lattes found himself enthralled by the case.

He soaked the bloodstains out of the cloth with distilled water, going to great pains to determine their precise weight. This might seem like over-zealous attention to detail, but he was determined to avoid 'pseudo-agglutination', where clumping occurs because the serum is too strong or because there are too many red blood cells in a test solution.

Despite the age of the stains, Lattes was still able to manufacture a few drops of liquid blood. This was then placed into small 'wells' on dimpled microscope slides, and drops of type A and type B blood introduced. The unknown blood agglutinated with type B, meaning that it must itself be type A. Therefore it was highly likely that it was the blood of either

Girardi or the lady visitor. Further examination under a microscope showed none of the epithelial (skin or mucus) cells that would be present in menstrual blood, demonstrating that it could not be the menstrual blood of the visiting friend. It was then determined that Girardi suffered from occasional bleeding brought on by a problem with his prostate. The weight of evidence was firmly on Girardi's side. His wife was forced to admit that she had been mistaken in her accusation and as a result Girardi's life became considerably less stressful.

Restoring marital bliss to a home might not be as dramatic an outcome as bringing a violent criminal to justice in the way some other forensic breakthroughs have, but it was nonetheless a triumph for Lattes – he had demonstrated that it was possible to determine the blood group of stains even three months old. The fascination with blood analysis that he developed during the case afterwards led him to specialise in serology and its application to crime detection.

Lattes soon followed up his success with Girardi by proving the innocence of a man charged with murder; using the same technique as before, he demonstrated that the blood group found on the suspect's clothing actually matched the suspect's own group, and not that of the murder victims. He then went on to invent a greatly simplified method of testing – placing tiny flakes of the dried blood he wanted to test onto the microscope slide, adding fresh blood, and then placing another slide on top. The serum in the fresh blood would do all the work of dissolving the blood that was to be tested, eliminating the time-consuming process of making up a liquid sample beforehand. If the bloods were of different groups the same clumping would still occur. Finally, in 1922, Lattes published his treatise on 'The Individuality of Blood', which went on to become a classic in the field. However, it was not until 1926 that he really became a household name in Italy (and indeed the rest of the world). It was in that year that he became involved in the Bruneri-Canella affair, a strange case that was to last for over forty years.

It began in a Jewish cemetery in Turin. A local caretaker witnessed a man acting suspiciously; at first he thought he was praying, but on closer examination realised that he was trying to steal a bronze funeral vase from a grave. When the man became aware he had been spotted, he took to his heels. He tried to conceal himself inside the church and then attempted to commit suicide, though the caretaker managed to apprehend him before he could succeed.

When questioned the man claimed to have lost his memory. Such was his behaviour that the local magistrate ordered that he be detained at the Collegno Mental Hospital, where he was listed as Unknown, Number 44170. The magistrate also arranged for the man's photograph to be printed in the local newspapers in the hope that someone might recognise him.

Sometime later, a woman by the name of Giulia Concetta Canella saw the photograph and became convinced that the man was her long-lost husband, Professor Giulio Canella. Canella had been a teacher and scholar of philosophy. In 1909 he had co-founded the *Rivista di filosofia neoscolastica* (*Neoscholastic Philosophy Review*) and in 1916 co-founded *Corriere del mattino* (*Morning Post*). He then married Giulia, who was his cousin and the daughter of a wealthy Brazilian businessman. They had two children together.

Canella had disappeared while on the Macedonian campaign

during the Great War. The only information his wife had about him was that during the battle of Monastir Hill he had been shot in the head and seriously wounded. According to his comrades, he had survived the injury but had been captured and made a prisoner of war. The fact that his body had not been recovered from the battlefield seemed to confirm this story. However, there was no record of him being taken prisoner, either.

Giulia asked to visit the asylum, and on 27 February 1927 she was granted a meeting with the unidentified man. Extreme care was taken – it was decided that to avoid putting the man under unnecessary stress the initial encounter should seem to be quite accidental. He was therefore taken for a walk through the cloisters of the hospital and things arranged so that he passed by Giulia Canella. But he showed not a hint of emotion or recognition when this occurred. The same could not be said of Canella, however – she insisted that there was no doubt whatsoever in her mind that the man she had seen was her husband.

As a result a second 'random' meeting was engineered. On this occasion the man told his attendants that he had a vague sense that he knew Giulia, and that he felt that his memory was stirring. On the strength of this a third meeting was arranged. This time Giulia, apparently unable to control herself any longer, broke down in tears. In response the man took her into his arms in a very familiar fashion. The doctors were finally convinced that the patient in their care was indeed Canella following a fourth encounter which occurred later that same afternoon, during which he talked about his recollections of his children.

Officially recognised as Professor Canella, in March 1927 the patient was sent back to Verona with his wife. Given the dramatic nature of the reunion the story unsurprisingly received a lot of press attention, with the Turin newspaper *La Stampa*, for example, running the headline 'A cry, a shiver, a hug, the light'.

On 3 March 1927, just a few days after the apparently happy ending, the quaestor of Turin received an unsigned letter stating that the man was not Canella at all, but rather a man called Mario Bruneri, a typist from Turin, born in 1886. He was an unscrupulous anarchist and conman who had been wanted by the Turin police since 1922 for acts of violence. He was also wanted in other cities, including Pavia and Milan. His extensive criminal record included theft and fraud, for both of which he had previously been incarcerated. Bruneri had been missing for six years, having apparently fled his family in order to live with his mistress.

The records held on Bruneri were quite wide-ranging and included a detailed physical and psychological profile, which perfectly matched the character and aspect of the man now claiming to be Canella. On Sunday 6 March 1927, the quaestor, firmly convinced he had been duped, arranged for the arrest of the man. He was brought back to Turin the same day.

Two days later, Bruneri's relatives were called in to see if they could identify him. His wife, Rosa Negro, was first. She recognised him at once, as did their fourteen-year-old son, Giuseppino, who ran up to him calling out, 'Papa, Papa!' Canella replied, 'Go little one and find your family as I have found mine.' When asked, 'Why deny that your son recognises you?' he replied with a wink, 'It's not for the son to recognise the father, but for the father to recognise the son.' His sisters Maria and Matilda, and his brother Felice, also recognised him and confirmed his identity.

But even in the face of this overwhelming evidence, the man refused to admit that he was Bruneri or to show the smallest sign of recognition towards his family. Even when one of his mistresses also recognised him, he stubbornly stuck to the story that he was Professor Canella. He even went so far as to fake a fainting fit in order to get out of the difficult situation.

The quaestor ordered fingerprints to be taken from the man, with a view to comparing them with Bruneri's criminal record. They were sent to the central police archive in Rome. While initially no match could be found, a second, more intensive search proved successful and the Scientific Investigation School of Rome wired back a telegram confirming that Bruneri and the man claiming to be Professor Canella were one and the same person. On the basis of this information, Canella/Bruneri was jailed in the Collegno Mental Hospital while awaiting trial.

It was now that Professor Lattes became involved. He pointed out that the truth could quite easily be established over and above the fingerprint evidence simply by comparing 'Canella's' blood with that of his parents and children. Blood groups are hereditary; if, say, both parents turned out to be group A, and 'Canella' was group B, then he certainly could not be their son. Equally, if he were group O and the children were group A or B, then it was impossible for him to be their father. 'I need merely one tiny drop of the blood of each individual concerned, and I can almost certainly establish beyond all doubt whether "Canella" is really the thief Bruneri,' Lattes asserted.

However, Lattes was not to be given the chance to test his theory – both 'Canella' and the family refused to give blood. The case dragged on, leaving opinion in Italy still divided about the man's true identity. He died on 12 December 1941, still claiming to be Professor Canella. Although blood analysis was not, in the end, used to establish the truth in this case, the large amount of publicity that it received still helped to raise public awareness of serology and to enhance Lattes' own reputation.

It was the German serologist Fritz Schiff (1889–1940), working in Berlin, who made the next significant advance. Although he was well aware of Lattes' methods, having had his book, *The Individuality of Blood*, translated into German, he had some serious reservations about them. Lattes' theories depended on Landsteiner's discoveries. As we have seen earlier in the chapter, these relate to the way different blood types cause each other to agglutinate or 'clump'. Landsteiner identified antibodies in the serum of blood and antigens in the red blood cells. The agglutinogens (antigens) in red blood cells can be of two types: antigen A and antigen B. The corresponding agglutinins (antibodies) in the serum are called antibody *a* and antibody B. These are distributed in the various blood groups in the following way:

Group A: Antigen A in the cells, antibody B in the serum.
Group B: Antigen B in the cells, antibody *a* in the serum.
Group O: No antigens in the cells, antibodies *a* and B in the serum.
Group AB: Antigens A and B in the cells, no antibody *a* or B in the serum.

So the antigens in cells of group A show clumping when mixed with the antibody in the serum of group B, because the group B serum contains its corresponding antibody. The antibody therefore attaches itself to the antigen, creating the clumping effect.

They also clump in O serum, since this also contains antibody *a*. But group O blood cells can be mixed with all three other serums without clumping, as they contain no antigens of either type. Group AB can receive blood groups A, B or O, as it contains no antibodies in its serum, and cannot therefore attach to any antigen. However, if group AB were to be given to either blood type A or B, clumping would occur, as the antibodies in both serums A and B would attach themselves to their corresponding antigen found in the AB cells. Given all this, it was theoretically easy for a serologist to determine the blood group of an unknown sample by a simple process of elimination.

Unfortunately, in group O blood, the B antibody loses its strength much more quickly than the *a* antibody. When this happened, the test might easily mistake group O blood for group B. Equally, the B antibody could degrade and vanish from a group A sample, making it appear to be type AB. The realisation that this complication could occur cast severe doubt on the reliability of Lattes' system of blood testing.

However, Schiff saw a potential solution. Although the antibodies in serum degrade, the antigens present in red blood cells retain their strength. Schiff theorised that this meant that if the cells from an old bloodstain were added to a fresh serum, they ought to produce some effect, even if they had lost their ability to agglutinate properly. They ought, in fact, to attract and absorb some of the serum's antibodies. Therefore, if a method of measuring exactly how much antibody was absorbed by the blood cells could be found, then the group of the old bloodstain could still be determined. It was a matter of measuring the effectiveness of the serum before and after the cells had been added to it. Having had this idea, Schiff worked hard at the problem but was unable to solve it himself. It was a young forensic scientist named Franz Josef Holzer who did that.

Holzer used dimpled microscope slides containing eight 'wells' in his investigations. He filled the wells with drops of group O serum (chosen because the serum contains both *a* and B antibodies so would react with both group A and group B cells), diluted to varying degrees with salt solution – each well contained a solution twice as dilute as the previous one. He then dropped exactly the same quantity of fresh blood cells into each well and observed how much each serum mix agglutinated them. Having recorded his findings, he then repeated the test, this time using an unknown bloodstain, and then rechecked each serum to see how far it had lost strength. Once again, it was then a simple matter of elimination.

A few years later, in 1934, the United Kingdom saw its first murder case involving forensic serology. Although the analysis of blood was not absolutely essential to the case, it nevertheless played an important role, and the bizarre and horrible nature of the crime means that the story is worth repeating. The pathologist involved was the noted Sir Sydney Smith (1883–1969), a New Zealander who had come to Britain to study medicine at Edinburgh. While there he became fascinated with the life and work of Dr Joseph Bell (1837–1911), who had lectured at Edinburgh University. He had been a pioneer in forensic science, and it was his incredible powers of observation and deduction that had inspired Conan Doyle in the creation of Sherlock Holmes. Through the application of Bell's methods for interpreting a crime scene, Smith had subsequently managed to unravel the complicated case of the death of a young officer in Egypt and to show that it was indeed suicide and not murder as had been suspected by some. By 1934 Smith was Regius Professor of Forensic Medicine at Edinburgh University. It was during this year that he achieved public recognition, for his work in connection with the murder of eight-year-old Helen Priestly.

Helen lived with her father and mother, John and Agnes, on the second floor of a drab and overcrowded tenement block, 61 Urquhart Street in Aberdeen. The flat comprised only two rooms, making for squalid and cramped living conditions. Helen was, by all accounts, a rather difficult child and prone to misbehaving.

On Saturday 21 April 1934, Helen's mother sent her out to buy some bread from the local Co-op, just a few hundred yards away. She arrived there safely and bought the bread, the baker noting the time of the sale as 1.30 P.M. However, after leaving the shop, Helen simply vanished. When it was realised that she was missing, a search was quickly organised. The streets and back alleys were scoured by local residents and the police. No trace of Helen was found anywhere.

It was then that a nine-year-old friend of Helen's, a boy called Dick Sutton, came forward with information that completely changed the shape of the investigation. Dick claimed that he had witnessed Helen being dragged down the street by a scruffy-looking man in a dark coat, who had then forced her into a tram. The police quickly circulated a description of the man and widened their search to the outer suburbs of Aberdeen. They also appealed for information on local radio and in local cinemas.

At 2.00 A.M. John Priestly and his friend and neighbour Alexander Parker returned home, both exhausted after searching long and hard for Helen. At 5.00 A.M., after just a few hours' sleep, Alexander decided that he would continue with the search, but that he would leave John Priestly to sleep a little longer. As he made his way downstairs, he noticed a large blue hessian sack stuffed under the stairs. Given the situation, he was a little suspicious and decided to investigate. When he opened the bag he made a horrible discovery: curled up inside was the body of Helen Priestly. It was later discovered that she had been strangled. Her pants were missing and there were bruises and other injuries on her thighs and genitals, indicating that she might have been raped.

Parker was questioned by the police. He was certain that the bag had not been there when he had returned home with Helen's father at 2.00 A.M. This led the police to believe that the murderer must have gone to Helen's home between 2.00 A.M. and 5.00 A.M. and left the body there to be discovered. However, it was soon realised that there was something wrong with this theory – during the night of the search it had rained heavily, yet the bag was still dry. So how did it get there? And how was it that, even with numerous people out on the streets for the search, nobody had been seen carrying the bag to the house?

The police re-interviewed young Dick Sutton in the hope of getting a better description of the man he had seen. He eventually admitted that he had made up the entire story, and had in fact seen nothing. Not only had his lies wasted hours of police time, they had also made them widen their search geographically when they should have been concentrating their enquiries much closer to home.

Given the mysterious appearance of the bag and the discovery that there was no 'scruffy man', the police were forced to reconsider the possibilities. Maybe, they began to think, the killer hadn't *returned* to Helen's block. Maybe they had never left it. They began to interview the local residents. Had the Priestlys been arguing? Had John or Agnes ever mistreated Helen? Was there someone within the building that might have wanted to hurt Helen for whatever reason? It was through following this line of enquiry that it was discovered that there was an ongoing dispute between the Donald family, who occupied the flat on the ground floor, and the Priestlys, who lived immediately above them. Alexander Donald was a hardworking barber while his wife Jeannie stayed at home running the household and looking after their young daughter, who was also called Jeannie.

Jeannie Donald senior had been seen arguing with Helen Priestly on a number of occasions, castigating her for her bad behaviour. Helen had certainly been known to provoke the Donalds; she apparently bullied their daughter, kicked at their front door, rattled the banister outside their flat and had even shouted abuse at Jeannie Donald. To get back to her flat, Helen would have to have passed the Donald family's door. Curiously, the Donalds were also the only residents of 61 Urquhart Street who had not participated in the search for Helen.

The police began to take a keen interest in the Donald family. At the same time, they concentrated their attentions on the hessian bag in which the body had been discovered.

There were several significant details for them to go on: it had been stamped with a Canadian export mark, it had once contained flour and held traces of washed cinders, an unusual cleaning method that was rather old-fashioned even then. It also had saucepan marks on it, presumably from being used as a makeshift tablecloth.

There weren't many places in the city that imported flour from Canada, but strangely enough one of the only ones was a bakery close to Urquhart Road. The police spoke with the owner, who confirmed that he had received a shipment of flour in exactly the same kind of sacks; he also confirmed that a customer had asked if she could have some of them and that he had given her several. The description he gave of the woman sounded remarkably like Jeannie Donald.

The evidence was now beginning to mount up; one of the residents of 61 Urquhart Road reported having heard a child scream at about 1.30 P.M. on the day of Helen's disappearance, a report confirmed by a slater who had been working in the alleyway behind the block.

The police decided to search the Donalds' flat. They discovered nine more bags identical to the one Helen's body had been found in, each one with similar saucepan stains. The most significant evidence, however, was the small bloodstains found on linoleum, newspaper, washing cloths and a scrubbing brush in the flat. It was now that Sir Sydney Smith and the forensic techniques at his disposal became important to the case. When Smith tested the blood it was found to be group O, the same type as Helen Priestly. This alone would have been damning evidence, but additionally Sir Sydney had discovered that Helen suffered from an unusual condition that enlarged her thalamus and caused her to produce a rare bacterium. Microbiological tests found this bacterium all over the Donalds' household: on the floor, on counter tops and on cleaning rags. Finally Sir Sydney examined the fibres of the bag and found that they contained cotton, wool, silk, cat hair, rabbit hair and some human hair that showed indications of having been badly permed.

The Donalds were arrested and interviewed. However, Alexander Donald was able to prove beyond doubt that he had been miles away at the time of Helen's murder and therefore could not possibly have been involved. He was subsequently released and the police turned their attentions to Jeannie Donald instead. Samples of her hair were taken and analysed by Professor John Glaister of Glasgow University. He was able to say with absolute certainty that Jeannie Donald's hair matched that found in the blue hessian sack. The evidence seemed incontrovertible: she had murdered little Helen Priestly.

It was obvious what the defence's main line of argument would be in a trial: that as a woman, Jeannie Donald was not capable of committing rape. To counter this, the prosecution had Sir Sydney carry out a further examination of Helen's body. One of the facts that had worried him during his initial examination was the complete lack of semen in or near the body. When he analysed the bruises and abrasions more closely, he came to the conclusion that they had not been caused by a rape but rather by the shaft of a hammer or a broom handle, the chilling implication being that the injuries were carried out in a deliberate effort to make the motive for the murder appear to be sexual.

This final discovery, in conjunction with the weight of the other forensic evidence (including the blood analysis), meant that there was no chance that Jeannie Donald would be acquitted. She was sentenced to death, though this was later commuted to life imprisonment. She was released in 1944 and died in 1976 at the age of eighty-one, having never admitted why she had committed such a horrible crime.

But, despite continued successes, it was often still difficult to convince a confused and not necessarily well-versed public (and crucially, therefore, juries) of the worth of forensic evidence. Such evidence might add weight to a case, but without a confession, it was rarely enough to secure a conviction. This is not to say that it was not useful, of course; it was not unusual for forensic evidence to push a criminal into admitting their guilt. We might consider for example, the case of Yoshiki Hirai, a young Japanese girl who was found raped and murdered in Japan in 1928. The police soon had two suspects in custody, one of whom was a beggar with mental health problems, who soon confessed to assaulting and murdering Yoshiki. Without blood analysis, it is likely that his story would have been accepted and he would have been convicted of the crime. However, testing showed that Yoshiki's killer was blood group A, while the beggar was blood group O. The other suspect, a man named Iba Hoshi, was blood group A, however, and when confronted with this evidence he confessed to the crime. Had he not, the chances of getting a conviction would have evaporated, since there are thousands of men with blood group A – in reality all the test did was exculpate the beggar, not implicate Hoshi, but it gave the police enough leverage to extract a confession. This lack of absolute precision would continue to limit the usefulness of serology in forensic detection until the development of genetic fingerprinting much later. Blood analysis was often still an important piece of the puzzle, but in truth by the 1950s more cases were being solved through the evidence of fingerprints and fibre analysis than through serology.

However, it was determined that evidence could be gathered not only by analysing the source and composition of the blood left behind, but also by observing its position and pattern; how it splashed, dripped, splattered, dropped, and sprayed. To a trained eye this can demonstrate how a murder or attack might have unfolded, and can reveal details such as whether the victim tried to fight back or run away (see Plate 7). The first study of bloodstain patterns was by Eduard Piotrowski at the Institute of Forensic Medicine in Poland in the 1890s. Subsequently, in 1895, he published a scientific paper on the subject called 'Concerning the Origin, Shape, Direction and Distribution of the Bloodstains Following Head Wounds Caused by Blows'.

One of the most celebrated cases in the annals of forensic science, which put exactly this kind of analysis to the test, is that of Dr Samuel Holmes Sheppard. It became one of the most infamous and controversial murder investigations in American criminal history.

Sheppard was born in Cleveland in 1923, the youngest of three brothers. He attended Cleveland Heights High School where he was an excellent student, holding the position of class president for three years. When he left school he decided to pursue a career in osteopathic medicine and enrolled at Hanover College, Indiana, before moving to the Los Angeles School of Physicians and Surgeons where he finished his education. In February 1945 he married his fiancée Marilyn Reese and together the two moved to a house in Bay Village, Ohio, so that Sheppard could join his father's medical practice.

But this comfortable picture of a young couple settling down to begin a life together was rent asunder when, during the early hours of 4 July 1954, Sheppard's wife, Marilyn Sheppard, was beaten to death in the bedroom of their home. She was pregnant at the time.

Sheppard explained that, after a dinner party, he had fallen asleep in the sitting room, only to be roused some time later, believing he heard his wife call his name. He immediately dashed up the stairs and saw someone grappling with her. He was then knocked out by a blow to the head. On regaining consciousness he was confronted with the bloodstained body of his wife. He heard a noise coming from downstairs, and staggered down to find a bushy-haired man making his escape via the back door. Sheppard gave chase, and attacked the figure, but was knocked unconscious once more. When he came round he found himself lying on the shore of Lake Erie (which the house backed onto), with his feet in the water and his T-shirt missing.

In a state of confusion he returned to the house and, finding that Marilyn was dead, called his neighbour, Mayor John Spencer Houk, before collapsing onto the living room couch. Houk and his family arrived at Sheppard's house at around 5 A.M. Shortly afterwards the police were called. They arrived with the Cuyahoga County Coroner Samuel Gerber. Gerber recorded that Marilyn Sheppard was lying face up on the bed, covered in blood and wearing just her pyjama top. She had been beaten to death, struck no fewer than thirty-five times. Beneath her body a pillowcase was discovered with a bloodstain that appeared to be in the shape of a surgical instrument, though it was not possible to ascertain what sort exactly. The house had been ransacked, though strangely the intruder had taken nothing, even ignoring several hundred dollars.

Gerber didn't believe Sheppard's story; he became convinced that he had murdered his wife during an argument and then staged a break-in to cover his tracks. The medical evidence pointed to Marilyn having been murdered at approximately 4 A.M., yet at least an hour had passed before Sheppard called for help, giving him plenty of time to dispose of incriminating evidence. As a result of Gerber's suspicions, Sheppard's story was probed more deeply. If Marilyn Sheppard's attacker was a burglar who simply wanted to silence her, why would he then go on to hit her thirty-five times, continuing his assault when she was already unconscious? Why would the intruder have remained in the house after his first run-in with Sheppard?

Additionally, the Sheppards owned a dog, yet it had not been heard barking when the house was supposedly broken into. Furthermore, Sheppard said that after his second encounter with the intruder he had woken up on the beach with his feet in the water, yet there was no sand in his hair, though admittedly his clothes – which were collected by Gerber – were wet. Finally, his bloodstained watch was discovered in a plastic bag in the garden of the house.

Six days after the murder, the police confronted Sheppard with yet another seemingly incriminating piece of intelligence when they asked him if he was having an affair with a pretty hospital technician called Susan Hayes. Nancy Ahern, one of the friends who had visited the Sheppards on the night of the murder, had given the police this information. Sheppard denied the accusation but, as more rumours and facts about the case began to circulate, the press started to turn against him. It was trial by media. At the inquest, Sheppard again stated that he had not had sexual relations with Susan Hayes. Unfortunately for him, Nancy Ahern gave evidence that Marilyn had told her that she knew her husband was having an affair with Susan Hayes, and that she was convinced that he was going to leave her. The grand jury met to consider the evidence and decided that there was a case against Sheppard. He was arrested and charged with murder. It took eighteen ballots to find a jury, but on 21 December 1954, Sheppard was found guilty of second-degree murder. However, this was far from being the end of the case.

Sheppard's defence attorney Bill Corrigan was unhappy with the verdict and began his own investigation into the evidence. He contacted Dr Paul Leland Kirk at the University of California, Berkeley. Dr Kirk was an esteemed criminalist who specialised in microscopy and was appointed the leader of UC Berkeley's criminology programme in 1937. He agreed to come to Bay Village and examine the evidence. His conclusions were vastly different from those of the prosecution.

It being eight months after the crime, Kirk was unable to examine a fresh scene. Instead he focused on recreating the crime based on the patterns of blood spatters in the bedroom. There were blood spots on two of the walls which he judged to have been caused by the battering of Marilyn Sheppard's head, while those on a third wall appeared to be spatter marks from a swinging weapon. The coroner Gerber had previously stated during the trial that he believed Marilyn had been killed by a surgical instrument. Kirk, on the other hand, concluded that the blood splatter from the weapon was almost certainly caused by a heavy object of some sort, such as a torch, and certainly not by a surgical instrument. He contended that the stain discovered on the pillow had in fact been made by the pillow being folded while it was still wet with Marilyn Sheppard's blood, and was not caused by any such instrument. Considering the amount of blood everywhere else, the killer must have been drenched, yet there was no blood found anywhere on Sheppard or his clothing. Kirk also determined that the killer must have held the murder weapon in his left hand, but Sheppard was right-handed.

Even with this new evidence, it took nine years to get Sheppard a retrial. It began in 1966. This time the well-known attorney F. Lee Bailey defended him. Bailey was a master of his art and quickly began to break down the prosecution case. He pointed out that the police search of the scene had been less than professional and that they had not tried to get fingerprints from several important items relating to the case. He then accused Gerber of being out to get Sheppard because he was jealous of him. Gerber lost his cool and responded angrily, something that didn't go down well with the jury.

Finally, Bailey brought everyone's attention to the blood spatters, particularly those on Sheppard's watch. The face of the watch had blood specks on it, which Dr Kirk had conceded looked like the splatter of flying blood. This would certainly have been the case if Sheppard had battered Marilyn to death while wearing it. However, Kirk also stated that as the tail on the end of the splatter was not symmetrical, he could not be sure that the spots were indeed 'flying' blood spots. Bailey then showed images of the blood spots found on the inside of the wrist strap, which were the same shape as the blood spots on the outside of the watch. It would not have been possible for these blood splatters to match, or even for there to be a blood splatter on the inside of the watch strap, if Sheppard had worn it while committing the crime. The blood on the watch could therefore not be used as evidence that Sheppard killed his wife. This, together with Kirk's other observations, impressed the jury and Sheppard was found not guilty. Speaking in connection with the case, Kirk later said, 'No other type of investigation of blood will yield so much useful information as an analysis of the blood distribution patterns.'

It's certainly true that blood distribution patterns can help us to reconstruct the sequence of events in a violent crime with surprising accuracy. On the other hand, we may find it hard to agree entirely with Kirk's statement, since the various forms of blood testing that have been developed down the years have provided strong evidence to help tie particular individuals to a crime scene – which is arguably of more practical use in actually securing a conviction. And, with the development of DNA fingerprinting by Alec Jeffreys in September 1984, the science of serology would take its biggest leap forward, something which we will explore further in Chapter 7.

4

Trace Evidence

Wherever he steps, whatever he touches, whatever he leaves, even unconsciously, will serve as a silent witness against him. Not only his fingerprints or his footprints, but his hair, the fibers from his clothes, the glass he breaks, the tool mark he leaves, the paint he scratches . . .

Paul L. Kirk, *Crime Investigation: Physical Evidence and the Police Laboratory* (1953)

On 18 July 1909, a quiet Sunday afternoon, the body of a young woman was found in a Parisian apartment. Her legs were spread wide and her face had been beaten so badly that she was almost unrecognisable. Police had broken into the apartment having been alerted that blood was leaking down from it and dripping through the ceiling of the café below. The concierge of the building had tried to gain entrance himself, but was obstructed by the iron gate blocking the stairs, while the window at the rear proved to be locked tight.

The flat belonged to a man named Albert Oursel and, despite her injuries, the woman was soon identified as sixteen-year-old Germaine Bichon, Oursel's mistress. Given that the contents of the flat had been completely ransacked, but that the remnants of a meal were found on the table, the police deduced that an intruder had surprised Bichon while she was eating, killed her, then burgled the apartment before escaping. However, as the only exits from the apartment – the window and the staircase – were both secure, it was a mystery how the perpetrator could have escaped.

A postmortem of Bichon's body by medical examiner Victor Balthazard confirmed that she had died from the injuries she had sustained during the attack, and also that she was five months pregnant. What was more revealing, however, was what he found curled in the palm of Bichon's hand: light-coloured hair that he believed to have come from the head of a woman.

Given the youth of the victim and the brutality of the crime, the case was clearly going to be very high profile, so Octave Hamard, the then chief of the Sûreté, took personal charge of the investigation. He quickly came to his own conclusions about the circumstances surrounding the crime and became convinced that Oursel had returned to Paris and murdered Bichon because she was pregnant with his child and insisting that they get married. This theory gained weight when the cleaning woman told him that the couple had been arguing for months.

Locating Oursel became a priority. He ran a domestic servant agency from the flat during the week but generally left Bichon alone there at weekends, staying instead with his family at their country home at Flins-sur-Seine. As it happened, Hamard was saved a journey there as Oursel returned to Paris the very next day. When questioned he turned out to have a cast-iron alibi; he had been at church at the time Bichon was murdered and the priest and many members of the congregation would be able to bear witness to the fact. After leaving church he had spent the afternoon lunching with his family. Even leaving all this aside, upon meeting him, Hamard felt Oursel was an unlikely killer – he was a weak, nervous little man with a strong attachment to his mother. After interviewing Oursel's secretary, who had left the apartment with Oursel on the day of the murder, Hamard discovered that seven francs had been stolen from a cash drawer and a further thirty from a bureau. A gold watch chain had also been taken, as well as a Russian rouble made from gold which was worth about forty francs. A fair sum in total, but not much for the life of a young girl. It was back to square one. Over the following weeks, theories and suspects came and went, with the police no closer to catching the killer. Then Hamard got a break.

It was discovered that on the Sunday prior to the murder a woman giving her name as Madam Bosch had accosted three servant girls not far from the flat. She had claimed that Oursel owed her a considerable sum of money. She wanted one of the girls to accompany her to his flat and witness her claim and his response. They had all refused, not wanting to get involved in such a contentious situation and finding the woman's behaviour decidedly strange anyway. After some investigation, Hamard found out that Bosch was the married name of a woman formerly called Rosella Rousseau, who was Oursel's previous cleaner.

Hamard had Bosch brought in for questioning. She completely denied any involvement with the incident the servant girls had described, and unfortunately they were unable to identify her with certainty. While he therefore lacked any hard evidence, Hamard could not shake the feeling that something was wrong. He decided to dig a bit more deeply into Bosch's background. It didn't take him long to find out that she and her husband were heavily in debt, to the extent that they had been in danger of being thrown out of their home. The day after the murder, however, they had suddenly been able to pay the rent. A neighbour also mentioned that Bosch had told him that she was going with her husband to a local shop to sell something valuable. Hamard's detectives found the dealer in question, who told them that Bosch had wanted to sell him a gold coin but that he had become suspicious and refused to buy it. It seemed a strong possibility that this was the Russian rouble Oursel had reported as missing.

Bosch was brought into the police station again. Hamard was now more determined than ever to find irrefutable evidence that would connect her to the crime. At this point the strands of hair discovered by Balthazard in Bichon's hand finally came into play. While she was in the police station, samples of hair were taken from Bosch and handed over to Balthazard for comparison. It did not take long to establish, using a microscope, that her hair was the same colour and width – 0.07 mm – as that recovered from the scene. Although this wasn't absolute proof that the hair was hers, the chances of it having come from anyone else were extremely small (see Plate 8).

While conducting his examination of the hairs, Balthazard remembered another important detail. Some of the hair that he had taken from the scene had been in a clump with blood on one end. He was convinced that this had happened when Bichon had literally ripped the hair from her murderer. The following day he paid a visit to Bosch, who had been remanded in prison. He examined her head and before long he found what he was looking for: slightly to the right of one temple was a clearly defined area where a tuft of hair had been pulled out.

Confronted with this evidence, Bosch finally broke down and confessed to the murder. Growing desperate as her financial situation spiralled out of control, and facing eviction, she had decided to steal from Oursel, her former employer. She knew that at weekends he was away from the flat and that she would only have Bichon to worry about. Her initial plan was to get someone else to distract the girl while she gained entrance to the flat – that was the reason she had approached the servant girls on the street. She then planned on hiding in the flat until Bichon left before carrying out the theft. When they refused to help she knew she would have to chance doing things on her own. She hoped that she would be able to slip into the apartment quickly while Bichon was out and make her escape before she returned.

The first part of this plan went well and she gained entrance to the flat without incident. Unfortunately, however, she lingered too long and Bichon returned. Bosch managed to conceal herself in a cupboard and decided that she would have to remain there overnight and wait for another chance to plunder the flat. The following day she thought her opportunity had come; Bichon seemed to have left the flat for lunch. Bosch crept out carefully from inside the cupboard only to discover to her horror that she had been mistaken – Bichon was sitting at the dining-room table, eating. Understandably shocked by the appearance of an intruder, Bichon attacked, driving Bosch back into the kitchen. She picked up an axe in order to defend herself, but in the struggle Bosch managed to pull it from her grasp and struck her across the face with it. As the young girl went down, Bosch subjected her to a flurry of blows, causing horrific injuries. She then forced open the cash box before ransacking the wardrobe and making her escape using Bichon's key to open the locks and secure the door behind her.

At her trial in February 1910, Bosch withdrew her confession, claiming that the police had forced it from her. She also tried to plead self-defence, saying she was fighting for her own life. It did her no good. The jury found her guilty and she was sentenced to death.

We have already referred to Edmond Locard earlier in this book. This brilliant forensic scientist made a very simple statement: 'Every contact leaves a trace.' This expresses the simple fact (now usually known as Locard's Exchange Principle) that even the cleverest criminals will almost certainly leave some small trace of their presence at a scene, or else take some small trace from the scene away with them. It is an idea that lies at the very heart of forensic science even today, for it is through such traces that an individual can be linked to a crime scene. The conviction of Madame Bosch is an eloquent demonstration of this principle in action.

Of course the kind of evidence we are referring to in such cases may sometimes be very hard to detect; to the naked eye a lot of crime scenes appear spotless and devoid of clues. It is for this reason that the gradual emergence of the use of trace evidence in forensic detection goes hand in hand with the development of microscopes and other such instruments; it was these that made it possible to analyse crime scenes in minute and painstaking detail.

Although glass existed in the ancient world and was used decoratively by various peoples, such as the Egyptians, its optical properties were not much explored at this time. The Romans did note its ability to magnify – the philosopher Seneca described how, when looking through glass filled with water, writing underneath the glass appeared enlarged. However, it is believed this effect was attributed to the water rather than the glass. It was also the Romans who discovered that lenses could be used to concentrate the power of the sun's rays to the point that it was possible to start a fire. As a result of these two properties, the first lenses were often referred to as either 'magnifiers' or 'burning glasses'.

But although such lenses were interesting, it wasn't until much later, at the end of the thirteenth century, that a widespread practical use for them was discovered. It was then that the first ever pair of spectacles was created in Italy. The invention is generally acknowledged to have been the work of an Italian man named Salvino D'Armate in 1284, though some doubt still remains about who really deserves the credit.

The next major advance in the use of lenses was in around 1590, when the Dutch spectacle makers, father and son Hans and Zaccharias Janssen, began experimenting with lenses by spacing several out along the inside of a tube. They found when placing an object under a lens, and looking at it through the lenses in the tube, the object was greatly magnified. They had created the compound microscope (a microscope that uses more than one lens).

The Janssens later claimed to have invented both the microscope and the telescope, though who was ultimately responsible for the successful creation of either remains open to debate as both work in very similar ways. A microscope uses a short focal lens to magnify a close-range object, and then this image is viewed through a long focal-length lens in the eyepiece. A telescope, on the other hand, uses a long focal lens to magnify objects far away, and a short focal lens is used in the eyepiece to view the magnified image. A lens-maker named Hans Lippershey (*c.* 1570–*c.* 1619) lived yards apart from the Janssens in Middelberg and also claimed the credit for both inventions. As Lippershey was the first to apply for the design patent of the telescope, he is now usually attributed with its invention, while Janssen is credited with the invention of the single-lens and compound optical microscope. Aside from this there is no evidence to prove conclusively in favour of either claimant. There are a whole series of confusing and conflicting claims taken from the testimony of friends and family during the many investigations that took place over the issue. In any case, the date of invention for the microscope is commonly given as between 1590 and 1595.

Despite such significant developments in the use of lenses for magnification, it was not until the mid-seventeenth century that the microscope was used in Europe for in-depth scientific examination. One of the first examples is the publication of *The Fly's Eye* in 1644. This detailed study of the anatomy of insects using a microscope was by the Italian astronomer and priest Giambattista Odierna, who was self-educated in science.

However, it was another Italian, the physician and biologist Marcello Malpighi (1628–1694), who truly pioneered the use of microscopes in biological study. While examining the structure of the lungs under a microscope in 1661, he observed that the lining of the lungs contained balloon-shaped sacs (alveoli), which were in turn connected to branch-like structures leading from small arteries and veins (capillaries). This historic discovery not only explained how oxygen is moved from the lungs into the blood, but also founded the science of microscopic anatomy, a field in which Malpighi made many significant discoveries throughout his career.

Following this, one of the greatest contributions to microscopy came from Dutch microscopist Antonie van Leeuwenhoek (1632–1723). Leeuwenhoek's fascination in developing and improving microscope lenses eventually enabled him to examine objects as small as one millionth of a metre. In 1674 he was the first man to discover single-cell organisms (now called microorganisms), including bacteria and sperm. In 1684, his advancements in microscopy enabled him to publish the first accurate observation of red blood cells. His work was confirmed by philosopher Robert Hooke (1635–1703), whose earlier work *Micrographia*, published in 1665, was widely respected for its intricate and accurate drawings of insects and organisms seen through a microscope. Leeuwenhoek's development of the microscope laid the foundations for future microbiology as well as promoting the use of microscopes for scientific study from the late seventeenth century onwards.

Further improvements to the microscope came in 1893 when the German scientist August Köhler developed a method to better illuminate microscope samples. He wanted to produce high-quality photographs from a microscope but was hindered by the uneven distribution of light supplied by then-achievable methods of illumination such as gas lamps. To remedy this, he used a collector lens to focus an image of an illuminating lamp into the front focal plane of the microscope, producing an evenly illuminated field of view without optical glare. Köhler illumination is still a central process in contemporary light microscopy.

Eighteenth-century microscopes on display at the Musée des Arts et Métiers in Paris. August Köhler's methods of illumination were to radically improve the images obtainable from microscopes such as these. This technology increasingly found its way into contemporary biology and, by extension, forensic science.

In 1891, before Kohler had even invented his new method of illumination, microscopic techniques finally made their way into the field of criminal investigation with the publication of *Handbuch für Untersuchungsrichter, Polizeibeamte, Gendarmen* (*Handbook for Coroners, Police Officials, Military Policemen*). Written by Austrian professor and judge Hans Gross (1847–1915), this groundbreaking book laid out why examining evidence under a microscope could be a vital step in solving crime. As this was the first book to combine the two fields of microscopy and criminal investigation, Hans Gross is considered by many to be the father of criminalistics.

Gross said of the emerging field: 'A large part of a criminalist's work is nothing more than a battle against lies. He has to discover the truth and must fight the lie. He meets the lie at every step.' Of course, the trick is not only recognising a lie, but also proving that it is a lie. It is the forensic scientist's job to see through the criminal's 'cleverness' and get to the truth. At times this can be extraordinarily difficult, particularly if one is dealing with a criminal of great cunning.

Gross highlights such a case, which he came across in an old record, in which a young man was believed to have burnt down the house of a farmer. The young man was the prime suspect for the crime, as it was widely known that he resented the farmer and used to work at the mill opposite the farmer's home. However, having left his occupation nine months previously, he was nowhere near the farmhouse at the time of the incident. On examining the scene, the investigating officers found evidence that, while still employed at the mill, the young man had constructed a device to set fire to the farmer's home at a later time. First he had stretched a strong spring and cord across a skylight in the granary that faced the house. Securing the spring with pitch, he had then arranged flammable material and a magnifying glass underneath the cord. Nine months later, the lens caused the sun's rays to focus onto the flammable material and ignite it. This in turn ignited the cord, which snapped, causing flaming pitch to catapult from the skylight onto the farmer's house.

While the complexity of the contraption might make us suspect this story to be apocryphal, for Gross it was a prime example of it being the criminalist's role to examine the evidence, see through the criminal's ingenuity and expose the truth.

The reason that Gross's *Handbook for Coroners, Police Officials, Military Policemen* is such an important publication in the history of forensics is that it combined in one volume information from several different fields of knowledge, including psychology and science, and presented it in such a way as to maximise its usefulness for those involved in criminal investigation. The contents also reflected the latest emerging trends in detection; by this time the work of Bertillon had lost much appeal with criminologists outside of France. Bertillon himself had been somewhat disgraced over the theft of the *Mona Lisa* in 1911; a palm print was found and Bertillon had no way of discovering whose it was, despite, as it turned out, having taken the measurements of the thief (an unbalanced Italian named Vincenzo Perugia) some years before. Gross's book placed far more emphasis on fingerprinting, and additionally pointed to the value of analysis of dust, hairs, wood fibres and other kinds of trace evidence.

Gross went on to found the Institute of Criminalistics in 1912 (which later became the Institute of Criminology) as part of the University of Graz Law School. This was to be the first of many similar institutes opened all over the world, formalising Gross's conception of criminalistics as a discipline in its own right.

One case that Gross investigated serves as an excellent example of the value of gathering trace evidence for analysis under a microscope. While walking their dog by the sea one day, three little girls underwent a frightening ordeal. They were accosted by a man who first exposed himself to them, then carried out a serious sexual assault on them, placing his penis inside their underwear. When they returned home, terrified, they immediately informed their parents, who called the police. The police followed standard procedure and collected all the clothing that the girls had been wearing at the time. When examined, the underwear of all three girls was found to have semen stains inside the crotch.

The following day one of the girls was out walking with her mother and was shocked to see the man who had assaulted her. The girl pointed him out and he was quickly arrested. Semen stains were discovered in and around the flies of his trousers. This seemed to confirm the girls' story, but on its own was hardly conclusive proof of his guilt. However, a dog hair was also recovered from the trousers, as well as some coloured wool fibres. When these were examined under a microscope, the hair turned out to be an exact match for those from the girls' terrier and the fibres to be an exact match for the dresses that the girls had been wearing. The case was proven. This was not just a triumph for justice and for Gross, but also an eloquent demonstration of the power of the microscope in dealing with trace evidence.

An unlikely leader in the field of forensic microscopy was Georg Popp (1863–1941) from Frankfurt, who had originally trained as a chemist. Although experienced in microscopic techniques from his work in laboratories, it was only in 1900 that he gained his first taste of forensic microscopy, when a criminal investigator asked him to use his expertise to examine some evidence. It was the beginning of a lifelong fascination, and a career in criminology. In 1889 he even founded his own laboratory, the Institute of Forensic Chemistry and Microscopy, which dealt with toxicological scientific analysis for the purpose of criminal investigations, as well as with related areas.

Influenced by Gross's book, he was a firm believer in the importance of fingerprints, and in the practice of photographing them. He was even able to use his knowledge in this area – and in chemistry – to solve a crime in his own lab. During the course of a theft, the culprit had touched a piece of platinum. Popp exposed this to vapours of ammonium sulphohydrate which caused fingerprints to show up in black wherever the thief had touched. From these it was soon possible to identify the perpetrator as a man who used the laboratory on a regular basis.

Fingerprints were also involved in one of the first cases to bring Popp to the attention of the public, when he was instrumental in solving the murder of a piano dealer in Frankfurt – he discovered and photographed prints at the scene of the crime and then compared them with suspects until he got a match. However, the case that really made him famous revolved around trace evidence, not fingerprints.

In October 1904, a young woman named Eva Disch was found dead in a bean field in Frankfurt. The postmortem examination showed that she had been raped then strangled to death with a scarf. A stained handkerchief had been found at the scene. Popp examined it under a microscope, and found nasal mucus containing traces of coal, snuff and the mineral horneblende.

Using this evidence, a man by the name of Karl Laubach soon became the prime suspect. He was known to use snuff, and worked in a coal-burning gasworks as well as part time in a local gravel pit that contained a large quantity of horneblende. Popp examined Laubach's fingernails and found coal and grains of minerals including hornblende underneath them. On examining Laubach's trousers, he found further evidence. Soil samples taken from them revealed two layers; the sample from the lower layer, directly in contact with the cloth, contained minerals that matched those taken from the crime scene. The upper layer contained particles of crushed mica mineral, which matched soil samples taken from the path from the murder scene to Laubach's home. Popp concluded that Laubach had picked up the first layer of minerals from the crime scene, before the layer of mica was added on top of it on his way home. Once confronted with this evidence, Laubach confessed. One of the Frankfurt newspapers of the day carried the headline 'The Microscope as Detective' in homage to Popp's painstaking investigation.

Popp garnered further nationwide acclaim in 1908 when he was instrumental in bringing the murderer of a woman called Margarethe Filbert to justice. On 29 May that year, in Rockenhausen in Bavaria, an architect by the name of Seeberger reported Filbert – who was his housekeeper – missing. She had taken a train to a nearby village the previous afternoon in order to go on a walk through the beautiful valley of Falkenstein, which boasted the spectacle of a ruined castle. She had failed to return home that evening.

After several days of searching, the police discovered her headless corpse in the woods. The initial impression given by the gruesome scene was that it had been some kind of sex crime; Filbert was lying on her back with her legs apart and her skirts pushed up. However, a later postmortem showed no sign of sexual assault. It was also noted that her purse, hat and parasol were missing, meaning that a motive of robbery could not be ruled out. The pathologist who examined the body concluded that she had been strangled before having her head removed using a sharp knife. Hairs were discovered in her hands.

A local magistrate called Sohn, alarmed by the savagery of the crime and aware of Popp by reputation, travelled to Frankfurt to seek his guidance in the case. He quite rightly felt that the hairs might be a vital clue that would reveal the identity of the killer and asked Popp to examine the evidence. Unfortunately, Popp quickly ascertained that the hairs came from Filbert's own head. However, he was now intrigued by the case and offered to continue his investigation.

The main suspect was a local factory worker called Andreas Schlicher, who was also a poacher whom witnesses claimed they had seen near the field on the day of the crime. When questioned he became indignant, denying any involvement with the murder. When a pair of his trousers, his gun and its ammunition were discovered at the nearby castle, he claimed to have left them there the day before the murder, saying he often did this to avoid alerting people to his poaching. Bloodstains were found on the knees of the trousers. When Popp soaked them in salt water and carried out the Uhlenhuth test, the reaction of the serum indicated that it was human blood. Further spots of blood were found on Schlicher's jacket, though it seemed that he had tried to wash them out. The evidence against him was mounting up, though for the time being was still inconclusive.

Popp then became interested in the encrusted soil on Schlicher's shoes. It had been established that Schlicher's wife had cleaned his shoes the day before the murder and that he had not worn that particular pair since the day of the murder. Together with a geologist, Popp collected soil samples from the area near the murder, from the area where Schlicher's possessions were found, and from near his house. Popp found that the samples taken from the murder scene contained decomposed red sandstone, angular quartz, ferruginous clay and a little vegetation. In distinct contrast, the samples from Schlicher's home contained fragments of porphyry, milky quartz, and mica, as well as root fibres, weathered straw and leaves. The area around the house was also littered with greenish goose droppings. Finally, the sample from the area where Schlicher's possessions had been found contained brick dust, coal and pieces of cement that had fallen away from the crumbling castle walls.

Armed with this information, Popp examined Schlicher's shoes. He found the sole encrusted with a thick layer of soil. Given that the shoes had been cleaned and then not worn bar on the day of the murder, Popp reasoned that the soil could only have accumulated on that day and that therefore each layer would contain a sequence of deposits from where Schlicher had been on the day of the murder.

He carefully removed the layers one by one. The earliest layer, attached directly to the shoe, consisted of goose droppings. On top of this was a layer of grains of red sandstone, and on top of that a mixture of coal, brick dust and cement fragments. These were obviously comparable to the samples that Popp had taken from the various locations of significance to the case. Schlicher claimed to have been walking in his own fields that day, but no fragments of porphyry with milky quartz were found in the soil on the shoes, which would have been the case if he was telling the truth about this. On the other hand, it seemed clear that the goose droppings came from near his home,

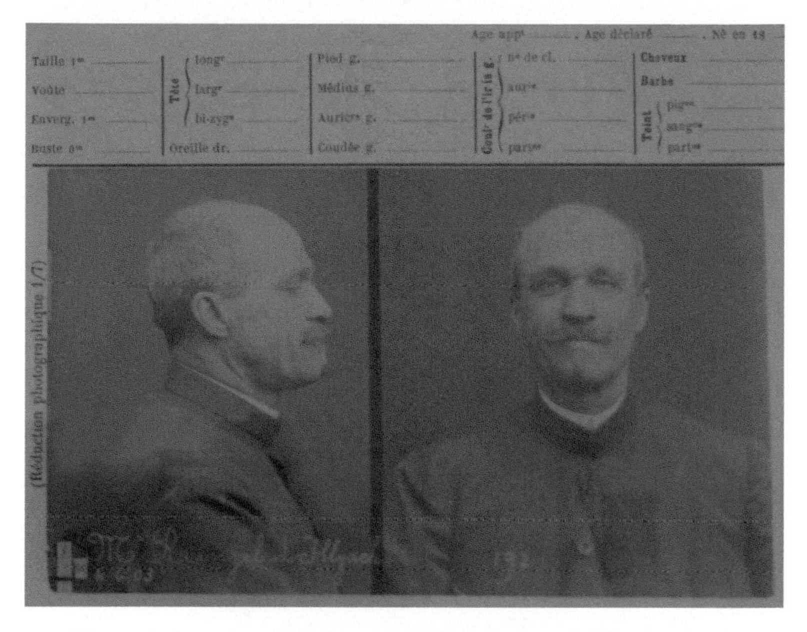

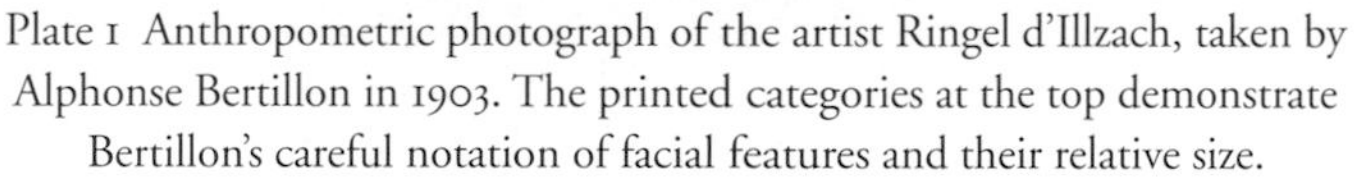

Plate 1 Anthropometric photograph of the artist Ringel d'Illzach, taken by Alphonse Bertillon in 1903. The printed categories at the top demonstrate Bertillon's careful notation of facial features and their relative size.

Plate 2 The cover of *Le Petit Journal* from 16th April 1892, illustrating the difficulty with which Ravachol was eventually arrested.

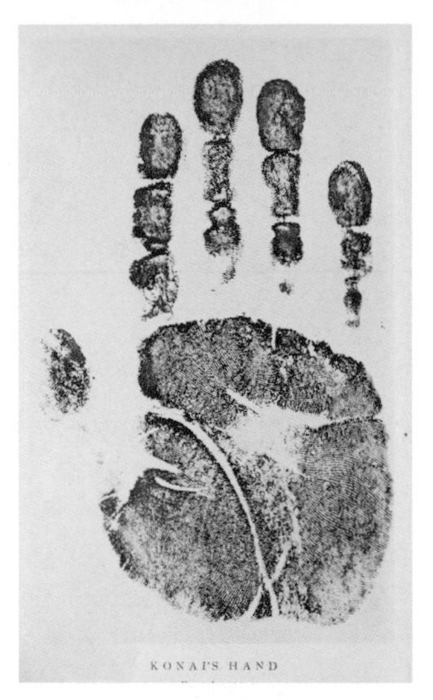

Plate 3 Sir William James Herschel (1833-1917) took hand prints from Bengali solders to try to prevent fraudulent pension withdrawals. This was instrumental in showing that fingerprinting could be used as a reliable means of establishing identity.

Plate 4 Detail from a 10th-century banner from Dunhuang which appears to show the first depiction of both 'fire-lances' and early grenades. Demons can be seen at the top right of the image, brandishing the weapons at a meditating Buddha.

Plate 5 A flintlock mechanism, showing both the spring-loaded cock and the frizzen which it struck, creating sparks to ignite the gunpowder.

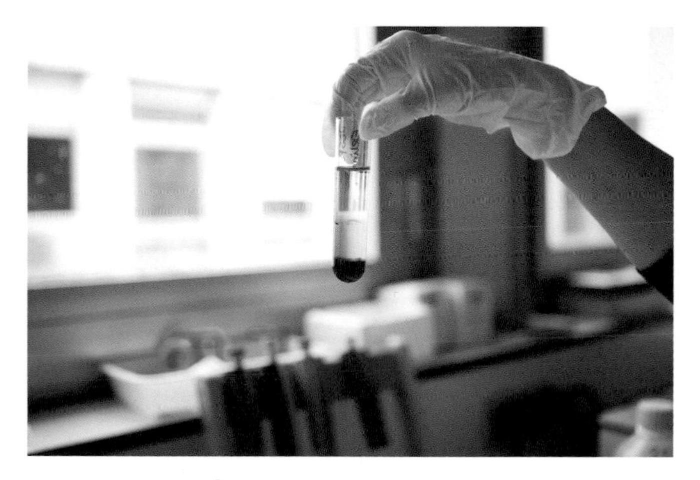

Plate 6 In the early twentieth century, blood analysis began to play a significant role in criminal investigations. Once Paul Uhlenhuth had devised a test to distinguish animal blood from human, killers like Ludwig Tessnow were no longer able to rely on the defence that bloodstains on their clothes were of animal origin. Tests for dried blood soon followed, and eventually tests to determine blood group. Blood analysis remains at the heart of many forensic investigations today.

Plate 7 The examination of blood spatter patterns can be extremely helpful in determining the events that took place during a violent crime. Dr Paul Leland Kirk once remarked that: 'No other type of investigation of blood will yield so much useful information as an analysis of the blood distribution patterns.'

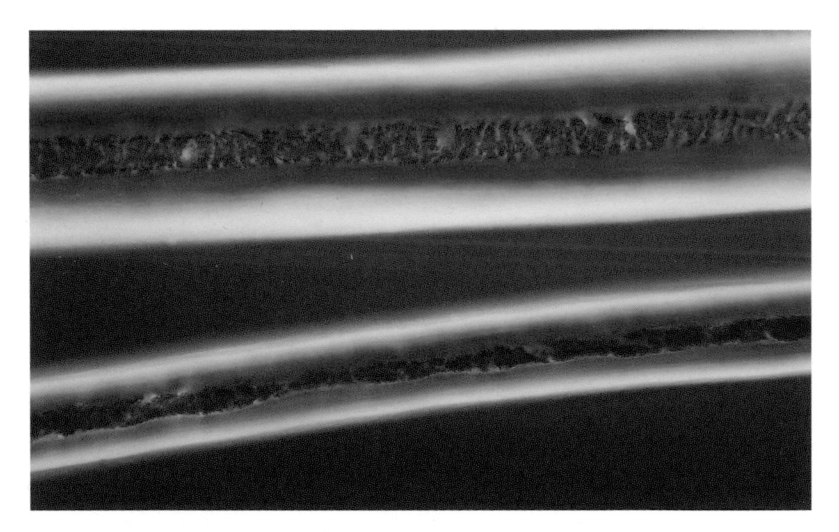

Plate 8 Two human hairs are compared underneath a microscope. As early as 1909, the close inspection of trace hairs on the body of Germaine Bichon was enough to establish that the suspected killer, Madam Bosch, had indeed been present at the scene.

Plate 9 By analysing the composition and layering of soil collected on shoes and clothes, forensic scientists are able to build up a complete picture of where a suspect or victim has been. These techniques were already being developed over a century ago; in the case of Margarethe Filbert's murder in 1908, Georg Popp used soil found on Andreas Schlicher's shoes to disprove his alibi, which led to Schlicher giving a full confession.

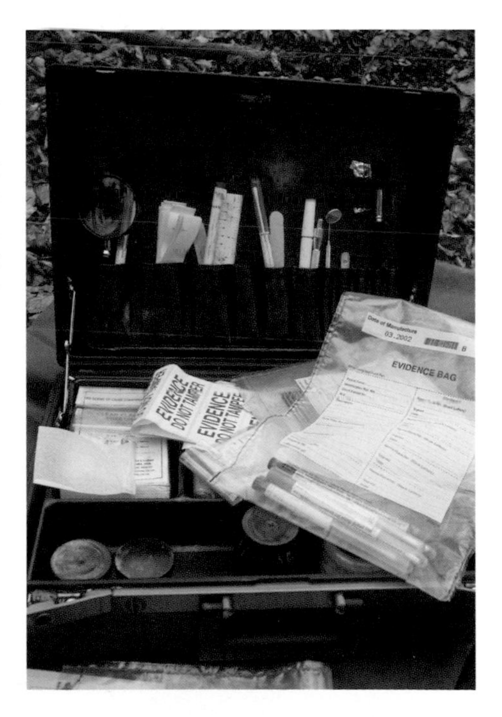

Plate 10 A forensic investigator's case. Bernard Spilsbury first identified the need for standardised crime-scene kits following the gruesome 'Murder at the Crumbles' in 1924. Today they are essential to any criminal investigation and contain a wide variety of equipment to allow evidence to be gathered safely and effectively.

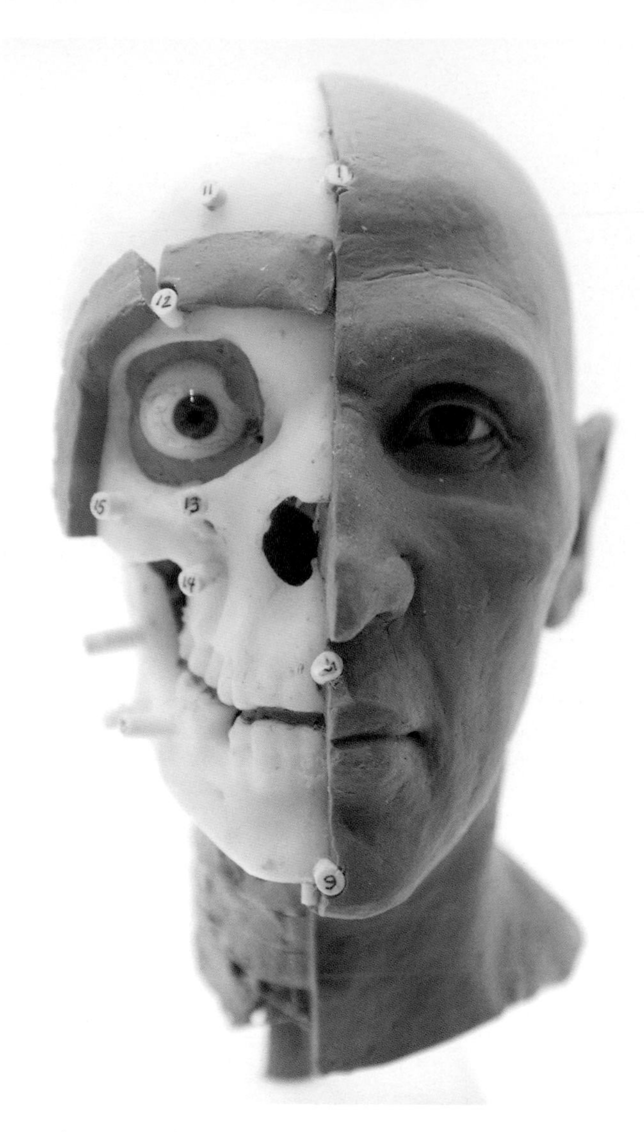

Plate 11 Facial reconstruction is now an important technique in establishing how decomposed remains might have looked when alive. The case of Buck Ruxton in 1935 employed similar methods, comparing the skulls of unidentified victims with known photographs of them to establish identity on the basis of similarity.

Plate 12 A newspaper's illustration of Marie Lafarge at the time of her trial in July 1840. She was the first person to be convicted almost entirely on direct forensic toxicological evidence, after she apparently poisoned her husband to escape an unsatisfactory marriage.

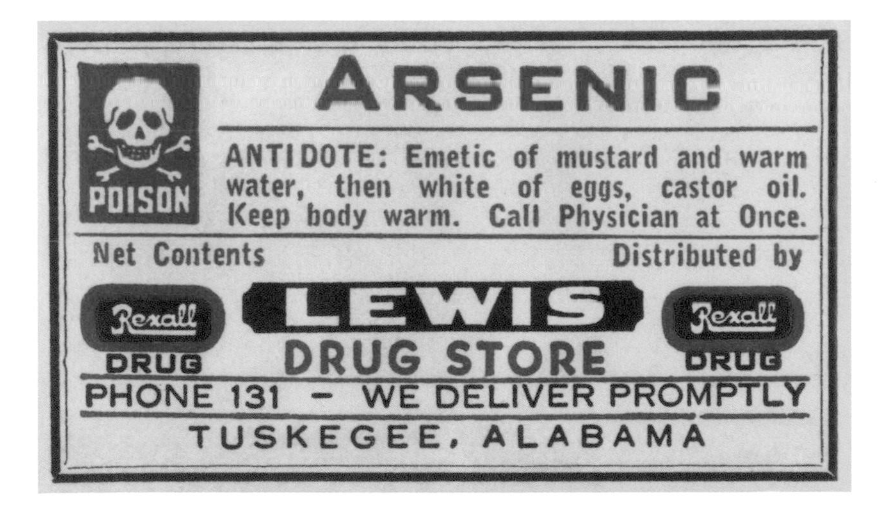

Plate 13 Label from a bottle of arsenic sold by a druggist in Alabama. In spite of its potentially lethal effects, for many years arsenic was available over the counter for a variety of household purposes.

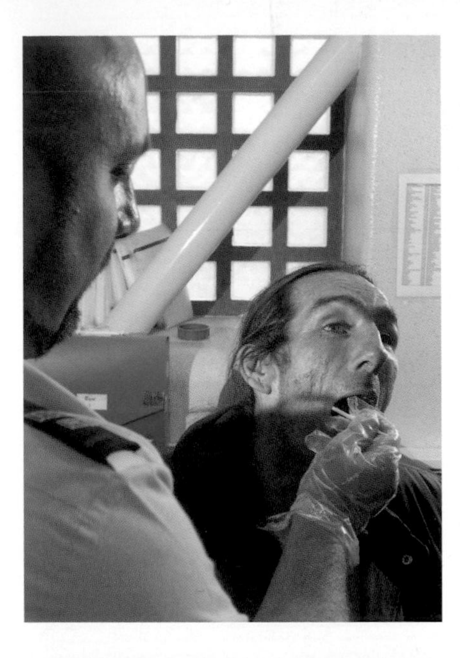

Plate 14 A cheek swab being carried out by a detention officer in order to gather a DNA sample. The advanced techniques used today mean that even a sample this small can be used to construct a DNA fingerprint which can then be compared against evidence found at a crime scene.

Plate 15 The Romanovs, the Russian imperial family, who were put to death in Yekaterinburg in 1918. When an unmarked mass grave was later discovered nearby it was widely assumed that the bodies it contained must be those of the family. However, it was not until the author approached the Russian Forensic Service in 1992 and offered to carry out genetic testing on the bodies that they were finally positively identified.

the red sandstone from the scene of the crime and the coal, brick dust and cement from the castle.

Faced with this compelling evidence against him, Schlicher finally confessed to the murder of Margarethe Filbert. From her appearance he had thought she was rich and decided to rob her. When he realised she had no money, he had attacked her in anger and cut her head off and hidden it. He was initially sentenced to death, though this was later commuted to life imprisonment.

The Margarethe Filbert case established Popp at the forefront of forensic geology and confirmed the vital role that soil samples play in criminal investigation (see Plate 9). The great Hans Gross had always maintained that the dirt on someone's shoes would eventually prove more compelling than a confession obtained by intensive interrogation. Popp had proved him right.

Microscopes continued to be promoted by other scientists as well. Professor Alexandre Lacassagne, the French physician and leading criminologist, impressed their usefulness upon his students at the Lyon Institute of Forensic Medicine. One of these students, Emile Villebrun, went on to become a leading forensic authority himself, specialising in fingernails; in the marks they leave and the value of material that might be discovered underneath them. He wrote a thesis on the subject and also solved a number of serious crimes. But perhaps the most famous of those who studied under Lacassagne is a man whose name we have already mentioned more than once: Edmond Locard.

Locard was born in Lyon in 1877. First educated at the Dominican College at Ouillins, he subsequently attended the University of Lyon, from which he graduated with a doctorate in medicine and a licentiate in law. He had developed a passion for all things forensic from an early age, spending his childhood reading Arthur Conan Doyle's Sherlock Holmes. After gaining his doctorate in medicine he was fortunate enough to be taken under Lacassagne's wing as his assistant. While still studying under him, Locard became convinced that both his country, and forensic science as a discipline, needed a real laboratory of crime, a laboratory completely dedicated to examining criminal evidence. This was extremely ambitious, as the idea had been tried by others before, including the famous Bertillon, and had always been met with indifference and even hostility. But Locard was determined not to let the short-sightedness of doubters and critics hamper him and so, after much persuasion on his part, in 1910 the police department of Lyon allowed him to create the first police laboratory in the attic rooms of the Lyon courthouse. However, the department did not officially recognise Locard's laboratory until 1912. He needed to prove himself, and his opportunity arrived in 1911.

It was discovered that there was a counterfeit gang in operation in the local area, forging coins and using them to buy things. The police already had their suspects, but no proof of their guilt. They arrested them in any case but, despite being questioned at length, they refused to confess. Frustrated by their lack of success, the police eventually turned to Locard who saw this as the perfect opportunity to prove his worth.

He began by examining the suspects' clothing, carefully going over each garment with his magnifying glass and tweezers. During the course of this exercise he came across an unusual-looking fine dust in one of the men's trouser pockets. He gently removed samples of this and placed them onto a large sheet of clean, white paper. He also brushed out the man's shirtsleeves, collecting the resulting samples on a second sheet. He looked at these samples under a microscope and was delighted to find that under powerful magnification he could clearly observe that the dust contained minute traces of metal. Chemical tests then revealed these to be tin, antimony and lead, which matched the composition of the counterfeit coins. Furthermore, he found that their quantities were in the same ratio. Similar evidence was subsequently recovered from the clothing of two more of the gang. When the evidence was presented to the suspects, they confessed.

The case was an enormous boost for Locard's reputation, and that of his laboratory. He had proved to the police that methodical scientific techniques could be of real practical value in crime detection. From that day on, his crime lab was never underused again and he went on to be instrumental in solving a great many crimes. Some years later, in 1922, he outlined some of the more infamous of these in his book *Policiers de roman et policiers de laboratoire* (*Detectives in Novels and Detectives in the Laboratory*).

One of the cases mentioned in the book occurred in 1912, when Locard was involved in investigating the murder of a young woman called Marie Latelle who had been found strangled one morning in the parlour of her parents' house just outside Lyon. The police were immediately given the name of her boyfriend, Emile Gourbin, as a potential suspect. Latelle was a pretty girl who apparently enjoyed flirting with other men, a habit that infuriated Gourbin. He had reportedly flown into a jealous rage with her over it on more than one occasion.

But while he seemed to have a possible motive, Gourbin also had an excellent alibi. A doctor who had examined Latelle's body had estimated her time of death at about midnight. On the night of the murder, Gourbin had spent the evening at the house of a friend who lived many miles from Marie's house. After passing the time eating, drinking wine and playing cards with several friends, he finally retired to bed at around 1 A.M. His friends confirmed his story; he was nowhere near her house on the night that Latelle was murdered.

The local police were at a loss and asked for assistance from their colleagues in Lyon, who suggested Locard might be able to help. He agreed to offer his expertise and conducted a full examination of the body using a magnifying glass. There were marks around Latelle's throat that the local police had assumed were the marks of the murderer's fingers. In fact, they turned out to be scratches caused by the murderer's nails digging in because of the strength with which they were gripping. This gave Locard an idea and he asked to see Gourbin. He inspected his hands and was pleased to discover that he did not appear to have cleaned them properly in recent days. Locard scraped the matter from beneath the young man's nails and transferred the residue to a section of white paper.

Locard returned to the laboratory with this new evidence and began to scrutinise it. Under a microscope he was able to observe that the material recovered from underneath Gourbin's nails included epithelial tissue – skin and blood cells. While this was perhaps a little suspicious, it was by no means damning evidence in and of itself; the cells could have come from Gourbin scratching himself. However, Locard noticed something else mixed in with the epithelial cells, a granular dust composed of regular-shaped crystals. This turned out to be powdered rice, a highly significant discovery since in 1911 this was the basic constituent of face powder. In addition Locard found iron oxide, zinc oxide, bismuth and magnesium stearate, all of which were chemicals commonly used in the cosmetics industry. The skin under Gourbin's nails had been covered in pink face powder.

Under Locard's instructions, Latelle's room was searched. A box of face powder was discovered. It was of a sort made by a local chemist and proved to be composed of ingredients identical to the material found under Gourbin's nails.

Confronted with this evidence, Gourbin finally confessed to killing Marie, explaining that he had duped his friends by advancing their wall clock, enabling him to slip out of bed, kill Marie and still have an alibi. Without Locard's methodical examination of the trace evidence, it is almost certain that Gourbin's alibi would have held and that the murder would have gone unsolved.

At this time Bertillon was still presiding over the forensic science department in Paris. However, in 1929 he was succeeded by the distinguished chemist Gaston-Edmond Bayle. Bayle had made his name in scientific circles with his work on spectroscopic analysis, where a spectrum is studied in order to determine characteristics of its source (for example, looking at the optical spectrum of an incandescent body to determine its composition). He had joined the Parisian police department as a forensic chemist and physicist in January 1915, though it was a further nine years before he became involved in a case that really gave him the chance to shine.

On 8 June 1924, the body of a seventy-year-old man called

Louis Boulay – who had disappeared on 30 May – was discovered in the Bois de Boulogne wrapped in a sheet. Both his wallet and a gold watch he was known to possess were missing from his person, suggesting robbery as a probable motive.

Bayle was asked to assist with the case. He quickly established that Boulay had been killed from several blows to the head with some kind of blunt instrument. He then began a careful search for trace evidence. He was not disappointed. Brushing through the hair of the corpse produced a mixture of river sand and sawdust, subsequent analysis finding the latter to be composed of oak and pine. There were also traces of coal dust, not only in Boulay's hair but also on his shirt. By determining its precise density, Bayle was able to identify it as anthracite. He also found traces of stone dust that he determined came from a grinding wheel.

Bayle also discovered two pieces of yellow cardboard on Boulay's clothes, the fibres of which were made of straw. From the victim's hat he recovered yeast cultures of a sort that you would expect to find in a wine cellar. Finally and most remarkably he discovered two beetles, both of which lacked eyes, indicating that they were a species that lived in total darkness. The man had evidently been battered to the ground somewhere where he had picked up these materials, before being removed from the place of his murder and dumped.

At first the police struggled to find any leads. Boulay appeared to be a respectable family man and it was difficult to see how he could have found himself embroiled in trouble. However, a newspaper was then discovered in his office with the names of two horses circled in pencil: Libre Pirate and Star Sapphire. It seemed that Boulay was fond of gambling. In fact, the bets involved were small and would have caused him no financial problems if he had lost, but since this was all they had to go on, the police decided to follow up on it.

They visited all the known gambling haunts in the city, showing Boulay's photograph at each to see if anyone recognised him. At first they had no luck, but eventually the landlord in a bar not far from the Gare St Lazare recognised him as one of his customers known as Père Louis. It seemed that he was well known as a racing aficionado and was well liked by other customers, to the extent that he often acted as a bookie's runner, ferrying money and bets for people.

This was the breakthrough the police had been looking for. Runners might easily end up carrying large amounts of money on them, and it seemed that both the horses that Boulay had circled in his paper had won at long odds. An elderly man carrying considerable winnings would be a tempting target for an unscrupulous person. Now what was needed was to track down the bookmaker's at which Boulay had placed the bets, and from which he would have had to claim the winnings. In the event, this proved enormously difficult. For the next five months the police interviewed dozens of bookmakers, both legal and illegal. Every one of them denied knowing Boulay and the investigation stalled.

Then the police got their second break. A clerk working in Boulay's office remembered him getting a letter from one of the bookies he used, a man by the name of Tessier, offering him a 'dead cert'. When prompted with this information, the chief clerk in the office also remembered Boulay mentioning Tessier, saying that he was no longer going to deal with him, on account of being unhappy with his office, which was situated in an old cellar. Much to the delight of the police, he was even able to remember Boulay telling him that these premises were situated on Rue Mogador.

Furnished with this information, the police were quickly able to build a more complete picture. The man in question was Lazare Tessier, a concierge serving 30 Rue Mogador. He was known to be an illegal bookie and had, in fact, already been interviewed about the murder; he had claimed that he had not taken any bets since his arrest a year ago. Now that they knew of its existence, it did not take the police long to locate the cellar from which Tessier did business. Bayle was called in to examine it, and during a careful search removed several samples. Analysing these back at his lab, he found that he was able to match various materials exactly to those discovered on the body – the coal dust, river sand and sawdust. Bayle was delighted, his only slight disappointment being that nowhere in his search had he discovered any of the sightless beetles. However, given the strength of the evidence he had succeeded in compiling, this was hardly something to worry about.

Tessier was promptly arrested but, in spite of the now compelling evidence against him (it was even discovered that he had been in debt but had suddenly been able to pay off his creditors), he still denied all charges. Although the police case was already strong, Bayle decided to return to the cellar for one more look, this time bringing a powerful lamp with him. With its aid he discovered what seemed to be spots of blood close to an area that had recently been repainted. He also found a bloodstain at the foot of the stairs and was subsequently able to prove that this was of human origin using the Uhlenhuth test, which contradicted Tessier's immediate claim that it had come from a cat. Finally, another tenant in the building mentioned that he had allowed Tessier to use his cellar upon occasion, and that when he had complained about an unpleasant odour emanating from it, Tessier had told him that it was coming from the drains, and that he would have them fixed. However, this was later attributed to the smell of Boulay's body rotting in the basement. Bayle checked the drains and found the sightless beetles he had been looking for inside them.

Tessier went to trial still denying the murder, in spite of the extraordinary weight of forensic evidence against him. He was found guilty of manslaughter and sentenced to ten years' imprisonment, while Bayle was hailed throughout France for his incredible investigative powers. Ultimately, however, his career would end in tragedy.

In mid-September 1929, having only just taken up the running of the Forensic Science Department in Bertillon's stead, Bayle was asked to examine a document that had been used by a travelling salesman, Joseph-Emile Philipponet, to procure money from his landlord. Bayle looked over the document in his laboratory for some time before eventually concluding that it was a forgery. When Philipponet was told, he took the news badly. Three days later, he managed to gain access to Bayle's laboratory and shot him three times in the back. Bayle died immediately. When he was arrested Philipponet claimed: 'Monsieur Bayle committed an act of bad faith! My document was genuine! What I have done was worth the death of a father of five children!'

So ended the life of one of the finest exponents of forensic science; it was a great loss to criminal detection.

Across the Atlantic, America was not slow to realise the usefulness of the microscope and the enormous crime-solving potential of trace evidence. Probably the finest example of these techniques being successfully used in the United States is the case of the murder of New York novelist Nancy Titterton in 1936.

Thirty-four-year-old Titterton lived with her husband Lewis (an NBC executive) at 22 Beekman Place, in an area popular with New York's literary set. She was a well-respected book reviewer and was considered a promising novelist. On Good Friday 1936, two men from a local furniture upholsterer's climbed the stairs to the fourth-floor apartment carrying a couch that the Tittertons had sent to be repaired. On arrival they found the door unexpectedly open. The older of the two, Theodore Kruger, called out to announce their presence. When he received no reply, he tentatively entered the apartment, followed by his young assistant John Fiorenza. They found no one inside and, since it seemed a rather unusual state of affairs, began to check the other rooms.

Looking into one of the bedrooms they found the bed in a dishevelled state, with the bedspread and ripped underwear strewn across the floor. Moments later they realised that the bathroom light was on and the shower was running. They stood outside the door and called out repeatedly, but once again there was no response. By now they had the definite feeling that something was amiss. Slowly they opened the bathroom door and peered inside.

Nancy Titterton's naked body was lying face down in the

empty bath. She had been strangled with her own pyjama jacket, which was still wrapped tightly around her throat. Understandably appalled by the dreadful scene, Kruger turned off the shower and hastened to call the police.

Because of Titterton's relatively high profile and the importance of her husband, a sixty-five-strong team was formed to investigate the crime under the leadership of Assistant Chief Inspector John Lyons. Several significant discoveries were made almost at once. Traces of green paint were found on the counterpane, and mud on the carpet. The biggest clue, however, came when Titterton's body was lifted from the bath. A cleanly severed thirteen-inch length of cord was discovered beneath it. Given the bruises evident on Titterton's wrists and the torn underwear in the bedroom, it seemed clear that the crime was sexually motivated and that the cord had been used to tie her hands before she was raped.

After murdering her, the killer had cut the cord in order to take it away from the scene with him, thereby leaving as little evidence as possible. Such careful planning suggested that perhaps the killer wasn't new to this; leaving the body in the bath with the shower running was also a clear attempt to destroy forensic evidence. Luckily for the investigative team, in his hurry to flee, the killer had failed to notice the stray section of cord hidden beneath the body.

In spite of the fact that there were several evidential leads to follow up on, initial forensic analyses of the scene proved disappointing. The mud on the carpet was discovered to contain traces of lint of the sort commonly found in upholstery premises; it had clearly been brought in by the two delivery men. The traces of green paint were revealed to have come from a decorator's can – the outside of the building was being painted. Apparently there were four men involved in this work, but other tenants in the building confirmed that only one had been in on the day of Titterton's death, and that he had been working on another floor at the time at which the murder must have occurred.

Given these disappointments, the police began to pin a lot of hope on gleaning some information from the length of cord. They started by checking with all the cord manufacturers in the New York area, to see if it had come from any of them. This proved fruitless, and they were forced to expand the search to cover not only New York but a further three states. Tracing the cord might yet prove to be useful, but it was clearly going to take some time.

In the meantime John Lyons continued to ponder the case. There were a number of elements to it that puzzled him. Firstly, the fact that the cord had been brought to the scene clearly showed that the murder had been planned, yet nobody had seen anyone suspicious in or near the building on the day. Besides, Nancy Titterton was a nervous woman, and it therefore seemed unlikely that she would have let a complete stranger into the apartment. That meant either that the killer had broken into the apartment somehow and surprised her, or else that she had known them. Lyons was convinced that it was the latter – but who was it?

It was a discovery by Dr Alexander Gettler, a chemist from the city's toxicology department, which at last began to guide Lyons towards a solution. Gettler had subjected the bedclothes in the disarrayed bedroom to close scrutiny with a magnifying glass and had been rewarded with the discovery of a stiff white hair about half an inch long that he was unable to account for. Examination under a powerful microscope allowed him to determine that it was a horsehair of the sort used to stuff furniture. When it was compared to the hair used to upholster the settee that Kruger and Fiorenza had delivered on the day of the murder, it was found to be a clear match. This might not appear of particular significance; after all, there were bound to be a few bits of the horsehair from the settee in the flat. However, the hair was too heavy to have simply blown into the bedroom, meaning that it must have been carried there in some way.

Working on that basic principle of Locard's that 'every contact leaves a trace', Lyons pondered this development. It was of course possible that one of the detectives examining the scene had carried the hair into the room with them, but Lyons felt it was more likely that either Kruger or Fiorenza had done so, since they would have had a great deal more contact with the settee and the horsehair it was stuffed with. However, both had stated that they hadn't actually entered the room, only stood by the door looking in for a short time. What, Lyons began to imagine, if one or even both of them had visited the flat earlier that day? It was something of a long shot, but in the absence of any other evidence, he decided to explore this theory.

Lyons returned to the upholsterer's shop and re-interviewed Kruger, who maintained that he'd been at work in the shop all day up until the point when he had gone with Fiorenza to deliver the settee. Fiorenza himself had also been in the shop, but had only come in at about midday; he had told Kruger that he had been visiting his parole officer, having been caught

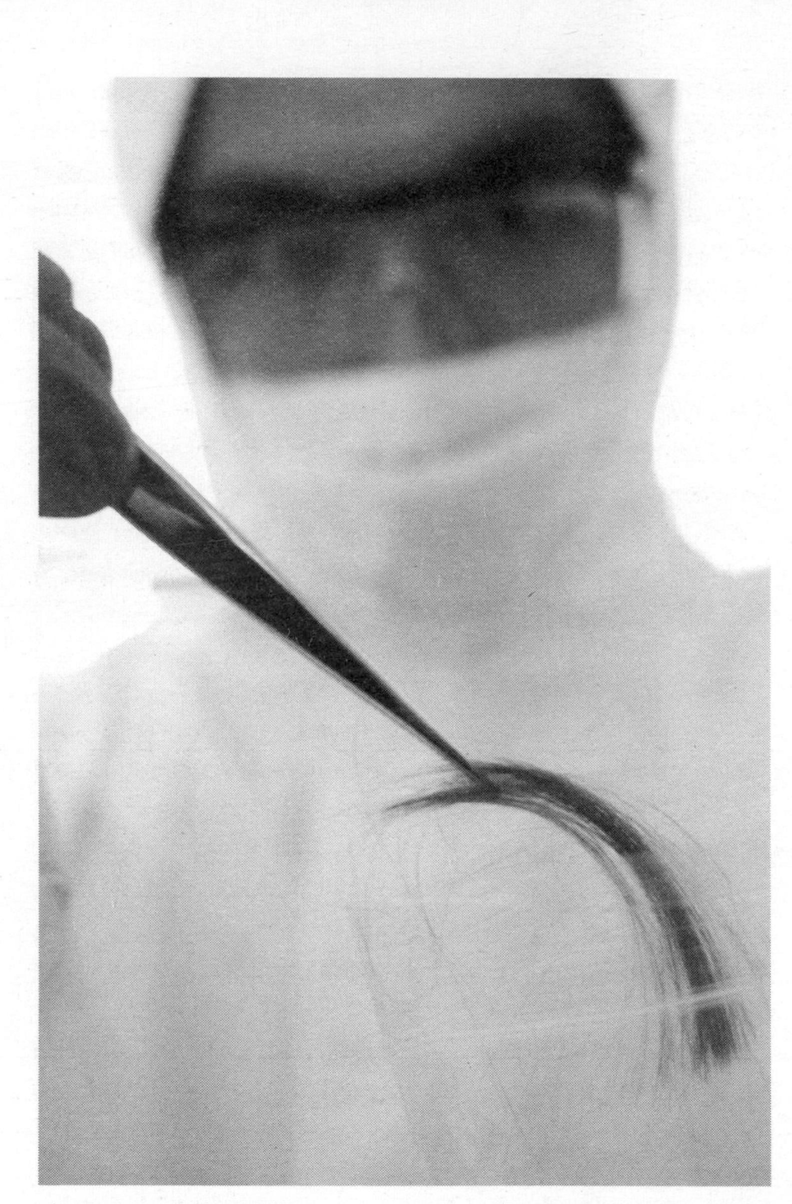

The examination of hair – often using a microscope to determine distinctive characteristics such as hair diameter – is now a key part of many forensic investigations.

stealing a car. Lyons also learned that Fiorenza had accompanied Kruger to pick up the settee in the first place, meaning that Nancy Titterton would have been familiar with him and might well have allowed him into the property, were he to have arrived with a plausible excuse for being there. Given this, when it was established that the parole office was actually closed on that particular day since it was Good Friday, Fiorenza became the prime suspect in the case.

Lyons had Fiorenza's criminal record looked into. He had been arrested four times for theft, and had spent two years in prison. More alarming, however, was a psychiatric report from 1934 in which Fiorenza was described as 'delusional' and 'prone to wild fantasies'. This certainly added further weight to Lyons's theory, but without any hard evidence to back it up, he felt arresting Fiorenza would be unproductive.

Then, on 17 April, came the breakthrough Lyons had been waiting for. The provenance of the cord found beneath Nancy Titterton's body finally came to light – it had been manufactured by the Hanover Cordage Company of York, Pennsylvania. Although it had been sold widely, Lyons's detectives were able to track down a wholesaler who had sold a roll to somewhere highly significant: the upholstery shop where Kruger and Fiorenza worked. It was the evidence Lyons needed.

Fiorenza was promptly arrested; although he initially denied having had anything to do with the murder, when confronted with the cord he finally broke and confessed. He explained that he had visited the apartment that morning on the pretence of returning the settee. When Nancy Titterton let him in, he attacked her. He dragged her into the bedroom and tied her hands with the cord. He then raped and strangled her before dragging her body into the bathroom and dumping it in the bath, where he 'found' her later when he returned to the apartment with Kruger.

Fiorenza went to the electric chair on 22 January 1937 for the premeditated murder of Nancy Titterton. He had been more aware of the potential for forensic evidence to betray him than many criminals, and by placing the body under a running shower and removing (as far as he knew) everything he had brought with him to the scene, he thought that he had concealed his guilt. But while a criminal in a hurry will almost certainly not be able to see every trace they have left behind at a scene, a meticulous forensic scientist can.

Fibre analysis also lay at the heart of a landmark case that took place in Liverpool, England. The night of 2 November 1940 was cold and wet. Fifteen-year-old Mary Hagen was therefore understandably reluctant to venture out when her father asked her to go and buy him a packet of cigarettes and a copy of the *Liverpool Echo*. However, at his insistence she pulled on her coat and left on the errand. She did not return.

The police were called and a search was mounted. Just five hours later, Mary's body was discovered in a cement blockhouse nearby. She had been raped and strangled. The evening edition of the *Liverpool Echo* lay next to her.

Amongst those called to the scene was Dr James Firth from the Home Office's Forensic Science Laboratory in the North West, situated in Preston, Lancashire. A scrap of muddy fabric had been found near the body and it was on this that Firth concentrated his attention. On close examination it seemed to be a bloodstained bandage. There was also a bloody thumbprint on the left side of Mary's neck. Since none of Mary's injuries had made her bleed, it seemed safe to assume that both the bandage and the blood had come from the killer; the bandage must have come off during the struggle and the injured thumb then pressed against Mary's neck as she was throttled.

When the bandage was analysed in the laboratory, an important discovery was made. The layer that would have come in contact with the wound was impregnated with a particular antiseptic called acriflavine and there were also traces of zinc ointment. The significance of this was that wartime prioritisation – remember this was 1940 – meant acriflavine was only commonly used for military dressings. Firth was therefore able to conclude that the killer was almost certainly a serviceman. This conclusion was supported by the evidence of a witness who had been asked the way to the local barracks by a soldier on the night of the murder. She had noticed at the time that his face was badly scratched.

As a result the police turned their attention to the nearby Royal Seaforth Barracks. It did not take them long to identify a suspect; a private in the Irish Guards by the name of Samuel Morgan had deserted some months before and was already under suspicion regarding an attack on another local woman called Anne McVitte, who had been robbed nearby. Morgan was returned to Liverpool and charged, though only with the attack on Anne McVitte.

The police, meanwhile, decided to investigate his family further. Morgan's sister-in-law admitted to harbouring him despite knowing that he was a deserter. When questioned she also said that she had treated the injury to his thumb using a dressing from Morgan's military kit, to which she had applied zinc ointment. She said he had told her that the injury had been caused by barbed wire. She still had some of both the bandage and the ointment left, and she promptly handed these over to the police.

What happened next was very unexpected. Morgan suddenly confessed to the murder – claiming he only intended to rob Mary – but denied rape. The investigating officers grew extremely suspicious; Morgan's confession along with the available evidence would be enough to land a conviction. But if he repealed his plea he might be able to get away. They needed more evidence, enough to convince a jury – with or without a confession.

Further forensic examination of the bandages eventually yielded the results the police were hoping for. When compared, the bandages that Morgan's sister-in-law had handed over exactly matched those found at the scene, as well as those from a third unopened packet found in Morgan's possession. However, military bandages were naturally prevalent across the country during the war, so this alone would not be enough to definitively tie Morgan to the crime scene. Firth began to look at Morgan's clothing, scraping the mud and soil from his clothes. He found that these samples contained traces of manganese, copper and lead. Examining samples from the blockhouse floor where Mary Hagen's body had been found, he soon discovered that these too contained manganese, copper and lead. This was strong evidence that Morgan had been at the blockhouse.

Firth then realised something else. While it was true that all the different samples of bandage that he had gathered were an overall match, those obtained at the scene and from Morgan's sister-in-law actually differed from the other samples. They had been double-stitched by hand, whereas the others had been single-stitched by machine. In fact, every other bandage from the barracks had been single-stitched by machine.

If Morgan's bandages had been exactly the same as all the others, his defence might have been able to argue that since there were thousands of such bandages belonging to thousands of soldiers, there was no way of proving that the one found at the scene came from the same set as those handed over by Morgan's sister-in-law; it might have been dropped by any soldier. However, Morgan had been unfortunate enough to pick up an atypical set of bandages from the barracks store, which very definitely placed him at the scene of the crime.

His trial began on 10 February 1941. As expected, he withdrew his confession, stating that the police had used coercion to obtain it. Nevertheless, his defence was unable to explain away the dossier of evidence that Firth had carefully compiled against him and he was hanged on 4 April 1941.

When they created the first microscope, those two Dutch lens-makers could not have known the impact their invention would have on the world, nor how vital it would prove to be in aiding criminal investigation. It opened up the possibility of detecting and analysing minuscule specks of material from a crime scene. Over the years the smallest pieces of trace evidence, even tiny particles of minerals or single hairs, have led to the resolution of some of the largest and most sensational criminal cases – proof, if any were needed, that Locard was right: 'Every contact leaves a trace.'

5

The Body

The human body is the best picture of the human soul.
Ludwig Wittgenstein

One thing that every murder has in common is that it leaves behind a body. On an emotional level the physical remains of a human being can be hard for a living person to face, particularly if they knew the deceased. While a corpse is clearly in some sense identical to the person who has died, and the violence done to a body is horribly evocative of their last moments, we are also keenly aware that there is something missing, that it is no longer fully a person. It is natural not to want to look closely at a dead body. However, as we have already seen in previous chapters, when studied in detail, a person's remains can yield a wealth of information.

When a body is discovered, the first job of the police is to try to identify it. As we have already seen, there are numerous ways to establish identity, though the most common today are for some form of ID to be found on the body, or for it to be recognised by a family member or friend. Of course the body also constitutes the single greatest piece of evidence that a crime has been committed, and since in the majority of murder cases the killer is known to the victim, establishing the identity of the body can often guide police straight to the perpetrator. Criminals will therefore often go to great lengths to hide, destroy or damage the body to prevent it being discovered and to ensure that, if it is, identifying it is extremely difficult.

However, human bodies are not easy to dispose of. They are difficult to burn or otherwise harm, with the bones and teeth particularly resistant to damage. They float in water, often even when weighted down. When hidden, decomposition quickly causes them to smell dreadful, and to draw the attention of insects and other wildlife. They are easily sniffed out by dogs being walked by their owners. And with continuing advances in forensic science, even remains that stay hidden and decomposing for many years, to the point of becoming completely skeletal, are still able to offer vital clues as to their identity.

The Parkman-Webster murder case remains one of the most sensational in American legal history, and provides a gruesome demonstration of the difficulties attendant on disposing of a body. It became a cause célèbre, largely on account of the horrible nature of the crime and the status of the people involved. It concerned the disappearance of a Bostonite, Dr George Parkman, on the afternoon of 23 November 1849.

Parkman was a wealthy socialite, businessman and philanthropist. He had a defined chin and wiry frame, and maintained a gentlemanly and restrained lifestyle. He worked hard, abstained from luxury, and was known for his amiable character. Fanny Longfellow, the poet's wife, labelled him 'the good-natured Don Quixote'. He was not without his enemies, however, with some viewing him as arrogant, money-grubbing and vain. In 1849 he was worth approximately half a million dollars, a considerable fortune.

The other important figure in the case was John White Webster. Webster had graduated from Harvard in 1811 and in 1814 was among the founders of the Linnaean Society of New England. He went on to graduate from Harvard Medical College in 1815, and after travelling extensively and marrying in 1818 returned to the college, having been appointed as a lecturer in chemistry, mineralogy and geology. He was then appointed as the Erving Professor in 1821, reflecting his reputation as a popular and well-respected academic. The notable physician and writer Oliver Wendell Holmes Sr was a dean and contemporary of Webster's at Harvard, and commented on his affable, if nervous, demeanour in lectures.

Webster was also known to have serious financial problems, which were in part the reason he sold off the family home in Cambridge, Massachusetts, in 1849, opting for a more modest, rented residence instead. In spite of this downsizing, his salary was still not nearly enough to cover his expenses. The books he had written, being academic in nature, made little extra money. He had been forced to borrow from several of his friends and was now struggling to pay these debts.

Parkman was among the people who had loaned Webster money. In 1842 Webster borrowed $400 from him. Five years later this was increased to $2,432, this sum including the original loan. As surety, Webster offered a cabinet of rare minerals. Unfortunately, by 1848 his situation had still not improved – in fact it had become worse. He had to borrow again, this time $1,200 from a man called Robert Shaw. In spite of the fact he had already offered it to Parkman a year previously, he once again used the cabinet of minerals as surety. Unfortunately, Parkman found out about this bit of deceit and decided to confront Webster.

So it was that on 22 November 1849, Parkman visited Cambridge in search of Webster. When he was unable to find him he called in on Mr Pettee, the Harvard cashier, and demanded that he give him all the money Webster had made from the sales of his lecture tickets to go towards repaying his debt. Pettee refused, since he was not authorised to pay anyone but Webster. The following day Webster, clearly disturbed at this turn of events, visited Parkman at his house and suggested that they should meet the following afternoon at the college in order to talk the matter over. Parkman agreed. The last time he was seen alive was at 1.45 P.M. on 24 November, entering the college along North Grove Street. He was wearing a frock coat, dark trousers, a purple satin waistcoat and his usual top hat.

When he failed to return home that day, Parkman's family reported him missing. Meanwhile Webster came home at 6 P.M. before attending a party that evening at the house of some friends, the Treadwells. He seemed in good humour, and certainly showed no outward signs of distress.

Two days later, on 26 November, Parkman's family offered a $3,000 reward for information leading to his being found. Twenty-eight thousand copies of a wanted notice were printed up and posted all over the district.

At this point one of the college janitors, a man called Ephraim Littlefield, began to play an important role in the story. Littlefield lived with his wife in the basement of the medical college, right next to Webster's laboratory. Some people had started to become suspicious of Littlefield and link him to Parkman's disappearance. Littlefield, in turn, was becoming suspicious of Webster's behaviour. He would later testify that on the day of Parkman's disappearance, he had heard the sound of running water coming from inside Webster's rooms, and found the door to them locked. He said that later that day he had seen the professor carrying some kind of bundle, and that Webster had asked him to make up a fire. Webster had also asked him a number of questions about the dissecting vault.

On 28 November, Webster arrived at the college early. Littlefield, alerted by his strange behaviour on previous days, kept track of his movements. Webster made no fewer than eight trips from the fuel closet to the furnace and back. The heat from the furnace became so great that the wall on the other side of it grew hot to the touch. When Webster left, Littlefield broke into his rooms through a window and discovered that even though he had only refilled them recently, all the kindling barrels were empty. He was now convinced that something was amiss, and determined to find out more.

On the following day, which happened to be Thanksgiving, Littlefield began to smash his way through the wall underneath Webster's privy. He had noticed that when police officers looked around the college as part of their enquiries, Webster seemed to draw their attention away from the privy. It was also an area that Webster alone had access to. Littlefield broke his way through two layers of brick before stopping, exhausted. The following day he resumed his work and managed to break through the wall completely. He crawled inside the space between the privy hole (which was several feet below) and the walls. It took a few moments for him to adjust to the darkness, but once he did he looked around. On top of a mound of dirt off to one side he caught sight of something out of the ordinary. He squinted at it, then, with a lurch of horror, realised what he was looking at: a human pelvis, a dismembered thigh and the lower part of a leg.

The police were called immediately and Marshal Francis Tukey attended the scene. The remains were removed from the room and laid out on a board, and the coroner, Jabez Pratt, was sent for. In the meantime Webster was arrested at his home in Cambridge. He denied all knowledge of the crime, expressing anger that the police would even consider him capable of such a brutal act. When they told him what Littlefield had discovered, however, he exclaimed, 'That villain! I am a ruined man!' He nevertheless then tried to blame the janitor for the crime, before later attempting to commit suicide by taking strychnine in his cell. The dose he took was not strong enough, though, and he succeeded only in making himself ill.

The police began the hunt for the rest of the body. When the sink in Webster's room was scrutinised it was found that it appeared to have been gouged in several places. There were also strange acid stains on the floor and steps of the furnace area. A button, some coins and some bone fragments, including a jawbone with teeth, were discovered inside the furnace. Finally a foul-smelling chest was found, also in the furnace area. It contained a hairy, armless, legless and headless torso. There was a thigh stuffed inside it, while the heart and other organs were missing. A right kidney and some blood-soaked clothing that belonged to Webster were later found elsewhere in the room.

The grim task of identification was left to Parkman's wife, who was able to confirm that the body was her husband's from markings near the penis and on the lower back. Later Dr Jeffries Wyman, the noted American naturalist, arrived to look at the bone fragments. He wrote a detailed report in which, amongst other things, he estimated the height of the man the bones came from to be approximately 5 feet 10 inches, a perfect match with Parkman.

Parkman was buried on 6 December 1849. It was one of the biggest funerals the state had ever seen, with thousands of people lining the streets. It is also a measure of the public attention the case had attracted that by this time over 5,000 'tourists' had also visited the crime scene. Webster's trial began on 19 March 1950 and ran until 1 April. It is estimated that, during that period, over 60,000 people attended the court, with tickets being handed out to the waiting crowds on a rotating basis. Journalists came from as far as London, Paris and Berlin to report on the story.

During the trial the defence argued that the body was not Parkman's and questioned whether a 'wound' discovered in the breast of the body had been a killing blow, since there was little blood near it – if it was not a killing blow, then there was no proof that anyone had actually murdered this person, whoever they were.

When Oliver Wendell Holmes Sr, the dean of Harvard Medical College, took the stand, he rebutted this, testifying that a wound between the ribs would not necessarily cause a great deal of blood loss. He also said that the body had been dismembered by someone with knowledge of anatomy and dissection, and noted that the build of the corpse was similar to that of Parkman.

The Parkman Murder.

TRIAL

OF

PROF. JOHN W. WEBSTER,

For the Murder of

DR. GEORGE PARKMAN,

November 23, 1849

Before the Supreme Judicial Court, in the City of Boston.

With Numerous Accurate Illustrations.

BOSTON:

PRINTED AT THE DAILY MAIL OFFICE,

14 & 16 State Street.

A booklet from 23 November 1849 detailing the judicial proceedings of the trial of John Webster for George Parkman's murder. The case captured the public imagination and drew a great deal of media attention.

Various other expert witnesses were called to the stand. Dr Charles Jackson gave evidence about the burning of corpses, noting that the 'furnace in the laboratory would have carried off the odour of burning flesh, if any had been consumed there'. Dr Jeffries Wyman demonstrated his findings from the bones and showed that they could be assembled into a whole. Parkman's dentist, Nathan Keep, weepily revealed to the court how the jawbone found precisely matched a plaster imprint he had kept of Parkman's jaw, and then showed how the loose teeth found in the furnace matched Parkman's plates: an inscription on the mould left no one in any doubt that Keep had made them for Parkman.

Despite attempts by the defence to rebut the evidence, on 1 April 1850 Webster was found guilty of murder and sentenced to death. On 4 May, his lawyers submitted a petition for a writ of error against Judge Shaw and his instructions to the jury. The writ was denied. Webster then appealed to Governor George N. Briggs for a pardon, asserting his innocence. This also failed and Briggs signed the death warrant.

Finally, in June, Webster wrote a confession, although he maintained that he had killed Parkman in self-defence after a fierce argument over his debt to him. He stated that Parkman was in such a rage that he thought he was going to attack him. In order to defend himself he had struck Parkman with a nearby length of wood, killing him. Webster was taken to Boston's Leverett Street Jail on 30 August 1850, and there publicly hanged. He was buried in the Copp's Hill Burying Ground. In an act of selfless compassion, Parkman's widow was the first contributor to a fund created for Webster's impoverished widow and daughters. Littlefield collected a $3,000 reward for providing information about Parkman's disappearance and was able to retire comfortably.

The Parkman-Webster case has been stitched into the scientific, cultural and legal fabric of American society. It stands as one of the first cases where the use of forensic science – and in particular dental evidence – led to the solving of a murder. The cultural impact of the case can be seen in the fact that grandees such as Charles Dickens, when visiting Massachusetts for the first time, insisted on being shown the room where Parkman met his fate. Historians have maintained an interest in the case into the twentieth century and beyond, and renowned historian Simon Schama centred his book *Dead Certainties* around it. There nevertheless remain doubts over the way the trial was handled. Some have said that Webster was dealt with unfairly by the judge; others claim that it was an unfair trial that nevertheless returned the correct outcome.

Although the Parkman case and others like it show the gradual emergence of a systematic approach to the treatment and identification of corpses, it is perhaps the French pathologist Jean Alexandre Eugene Lacassagne who deserves the most credit for transforming it into an exact science. Lacassagne was born in Cahors in 1843 and attended the military academy at Strasbourg before going on to qualify as a military surgeon. He became familiar with wounds of various kinds, including observing operations on gunshot wounds, during his time on campaign in North Africa. On leaving the army in 1878, he wrote *Précis de Médicine Judiciaire* (*A Summary of Judicial Medicine*), drawn from his experiences as an army physician. As a result of this, in 1880 he was invited to occupy the newly founded chair in forensic medicine at the University of Lyon. During his time as head of the department, his expertise was called upon by the authorities to help conclude a great variety of cases. He also coined the phrase 'One must know how to doubt', and drummed it into the heads of all his students.

Lacassagne's greatest case occurred in August 1889. A local official in Millery, a small town just south of Lyon, was asked to investigate a foul stench coming from near the river. With the help of several council workers he eventually traced it to a canvas bag, which had been dumped in some bushes. With rags pressed firmly to their noses – for the smell was truly appalling – they dragged the bag from its hiding place. The official then loosened the ties holding it together.

If the smell had been bad, the sight that confronted them was far worse. Inside were the naked and decomposing remains of a dark-haired man, wrapped in oilcloth and string. The police were summoned at once and the corpse was removed to the Lyon City Mortuary, a rotting old barge anchored in the middle of the River Rhône. A Dr Paul Bernard then conducted a postmortem: gruesome work. Due to the state of the body it was at first difficult to establish a cause of death, but he was eventually able to conclude that the unknown man had been strangled. Bernard estimated that he was about thirty-five years old.

A few days after the discovery of the body, a wooden trunk was also found. From the stink of rotting flesh that clung to it, it was assumed that the body had been contained inside it at some point. Although it had been in the water for some time, it still contained a useful clue: fragments of a railway label indicating that it had been sent to Lyon's Perrache train station from Paris on 27 July.

News of the macabre case made headlines all over France,

with other European newspapers also picking up the story. Assistant Superintendent Marie-François Goron of the Sûreté in Paris was put in charge of the case. Searching through the missing person files he came across a name, Toussaint-Augsent Gouffe, a forty-nine-year-old bailiff and a notorious philanderer. He had been reported missing on 27 July by his brother-in-law, Landry.

Gouffe lived on Rue Montmartre with his three daughters. His sex drive was legendary and he spent most nights searching the cafés and clubs of Paris for potential partners. On a Friday he would often stay out all night, having inveigled himself into the bed of a woman, leaving the day's takings in his office. The warden of the building in which the office was housed was therefore rather surprised when, at about 9 A.M. on 27 July (which was a Saturday), he heard him going up the stairs. When he heard him coming back down again a short while afterwards, he went to meet him in order to exchange a few pleasantries. However, it wasn't Gouffe at all, but a stranger, who immediately ran from the building. Thinking that it must have been a robbery, the caretaker went upstairs to check on the office, and was mystified to discover that Gouffe's takings of around 14,000 francs were still there.

To be sure that the victim was indeed Gouffe, Goron arranged for Landry to view the corpse. With so many decomposing bodies situated around the mortuary-barge, the smell was overwhelming; Landry glanced quickly at the remains before running outside and being violently ill. He also informed Goron that the body could not be that of his brother-in-law because the hair on the chest of the body was black and Gouffe had chestnut hair.

This was a disappointing setback for Goron but he was not a man to be easily dissuaded. He questioned Bernard again but the doctor confirmed that the corpse did indeed have black hair, not auburn. Still convinced that the body was Gouffe's, Goron asked Bernard to collect several strands of hair from the dead man's head. He then immersed them in distilled water. It did not take long for the water to wash away the outer coating of dust, blood and dirt that had become stuck to the hair, revealing that its true colour was indeed auburn. Bernard was both amazed and terribly embarrassed. In the time it took for this discovery to be made, the body had already been buried in a cemetery at La Guillotière, on account of its rapidly worsening state. Goron ordered it to be exhumed immediately and in mid-November it was delivered to Lacassagne's laboratory at the university. It was not unusual for Lacassagne to be called in when it seemed likely that other doctors might have overlooked something, and he had also been away when the body was originally discovered. He began his work.

By now the remains were in a truly terrible state. The genital organs had completely decomposed, most of the facial and body hair had disappeared and parts of the skull were missing. Lacassagne's first task was to scrape the outer layers of flesh from the remains – the decay of the body was so advanced that it would not have been possible to learn anything further from this.

Bernard had botched things with the initial autopsy. He had used a hammer instead of a saw to take off the top of the head, which meant that Lacassagne could not check for head trauma; he had destroyed the sternum with a chisel so chest trauma could not be found either; he had also left the body with bones out of place, and organs had been removed and put into a basket. Still, Lacassagne was able to observe that the right knee was deformed, and that parts of the bones there to which muscles attach were underdeveloped. He also found evidence of a tubercular infection of the leg during youth. Taken together, these things meant that the man had almost certainly walked with a limp. Having been given this information, Goron was quickly able to establish from Gouffe's relatives and his shoemaker that he did indeed walk with a limp.

Lacassagne also scrutinised the teeth of the body, after which he estimated the deceased to have been in his early fifties, certainly not thirty-five as Bernard had suggested after the original postmortem. Gouffe was, as we know, forty-nine. Breaks in the thyroid cartilage confirmed strangulation as the cause of death, although Lacassagne considered that it had been done manually rather than with a cord or a garrotte as Bernard had thought. The final test was the hair. Lacassagne used a microscope to compare some hair that had been taken from the corpse with some taken from Gouffe's hairbrush. They matched in colour and were both 0.13 mm in diameter. Both Lacassagne and Goron were now completely convinced that the remains were indeed those of Gouffe. Had it not been for Lacassagne's determined and systematic approach, it is very likely that this would not have come to light, especially given the inaccurate information arising from the first postmortem and the confusion caused by the filthy hair.

But while they now knew for certain who their victim was, they still needed to find a killer. In the hope of jogging someone's memory, Goron commissioned a copy of the trunk the body had been carried in, which was then put on public display. An estimated 35,000 people filed by to see it, and thousands of photographs were also circulated around the world. This might have seemed like a long shot, but it worked – Goron received a letter from a Frenchman now living in London, who had seen a photograph of the trunk in the paper. In it, the man explained that an ugly, balding man calling himself Michael, along with his daughter, had resided with him the previous June. They had bought a trunk just like the one in the photograph from a shop near Euston Road and had taken it with them when they finally returned to Paris.

The description of the man calling himself Michael fitted one that Goron had been given of a man called Michel Eyraud, with whom Gouffe had been seen drinking a couple of days before his disappearance. With them was Eyraud's beautiful mistress (not, in fact, his daughter) Gabrielle Bompard, a former prostitute. Given Gouffe's eye for attractive women, it seemed this might well be the lead that Goron had been looking for. When he received the information, Goron at once ordered a search for Eyraud and Gabrielle, but they had gone to ground.

Once again, Goron turned to the newspapers for help. Luckily they couldn't get enough of the story and relished the chance to become involved. Soon they were full of descriptions, drawings and background information, all pertaining to Eyraud and Bompard. Then, completely out of the blue, Goron received a letter from Eyraud stating that he was now living in New York and wondering why Goron was implicating him in the murder of his friend, of which he denied all knowledge. He went on to offer an alternative theory, mentioning Bompard as his former mistress and suggesting that perhaps she was in some way involved in the murder. He also promised to return to Paris and hand himself over to Goron.

The surprises didn't stop there. A few days later, Bompard appeared at the police station with her current lover in tow. Goron described her as being small and pretty, with grey eyes and excellent teeth. He later reflected that 'corruption literally oozed from her'. She had come to denounce Eyraud as Gouffe's murderer and admit that she had been his willing accomplice.

When Goron questioned her she was very frank and straightforward. The murder, she said, had been committed in one of the rooms at 3 Rue Tronson de Coudray. Although she said she knew this to be a fact from what Eyraud had told her, she claimed she hadn't actually been present at the time of the crime. She had then travelled to America with Eyraud. While there she left Eyraud for another man (the one she had brought with her to the police station), and Eyraud had apparently planned to murder and rob this usurper. However, she told him about Eyraud's scheme and together they escaped back to Paris. It was he who had then persuaded her to come and tell the police what she knew. Goron appreciated her bravery in coming forward but he nevertheless instructed the prefect, Loze, to arrest her and hold her in custody.

Goron sent men to search for Eyraud in America and Canada, but they were unable to catch up with him, despite tracking a long trail of his petty crimes, and eventually returned to France empty-handed. But one misdemeanour in particular eventually proved to be Eyraud's undoing. While in New York he 'borrowed' an expensive oriental robe from a Turkish gentleman, on the pretext that he wanted to be photographed wearing it. Needless to say the poor man never saw his robe, or Eyraud, again. He subsequently travelled to Havana, Cuba, and while there tried to sell the stolen robe to a dressmaker. In a stroke of luck for Goron and Lacassagne, the dressmaker recognised Eyraud from having seen his photograph in the paper, and promptly informed the French consul. The hunt was on.

Police raided Eyraud's room at the Hotel Roma in Havana. There they found his belongings packed and ready for a quick exit, but no sign of the man himself. Later that night he tried to gain entrance to a brothel, but the madam, suspicious of his ragged appearance, threw him out before calling the police. It did not take long for them to locate him wandering the streets and arrest him. They finally had their man.

On being brought back to Havana Police Station, Eyraud attempted and failed to commit suicide. He was then transported back to Paris where he confessed and told his version of the story. Contrary to Bompard's account, Eyraud said she was very much involved, and that he had persuaded her to lure Gouffe to a room where he would be lying in wait for him. Bompard would then begin to seduce him and, while he was distracted in this way, Eyraud would strike. Things went to plan; Eyraud attacked Gouffe, first attempting to hang him and then, when he started screaming, resorting to strangling him with his bare hands. The body was hidden inside the trunk. Eyraud then left in order to break into Gouffe's office to steal the takings he knew to be there – this was the entire motive for the crime. However, for some reason – most likely because he was in something of a panic – he was unable to locate the cash. Later he disposed of the body in the river, thinking that would be an end to the matter.

Had the two committed the murder in another area, one not served by Lacassagne, it is highly probable that they would have got away with it, since without him Gouffe's corpse would have remained unidentified. Unfortunately for them, they didn't. After a four-day trial, Eyraud was found guilty of murder and sent to the guillotine. Despite the important part she had played, Bompard was treated more leniently and sentenced to twenty years in prison.

The high profile of the case, both in France and internationally, considerably enhanced Lacassagne's own reputation, not to mention being a huge boost to that of forensic science as a whole.

The discovery of a corpse in a bag seems more the stuff of nightmare or Hollywood thriller than real life, and of course in truth such things thankfully only occur extremely rarely. However, a couple of decades after that unpleasant episode in France, New York had its own case of an unidentified corpse that would prove to be every bit as sensational. Although it was not forensic analysis of the body itself that solved this case, the circumstances of the crime and its peculiar nature merit its inclusion in this chapter.

On 13 September 1913, eighteen-year-old Mary Bann and her eleven-year-old brother Albert were looking out from the porch of their Palisades home overlooking the Hudson river. This was something they often did. On this occasion, however, they spotted a parcel being carried along on the early morning tide. Even as they watched, it came ashore. Their curiosity got the better of them and they hurried down to the edge of the water to find out what the package contained.

Pulling the manila paper apart, they discovered a red and blue striped pillow. It had been slit open and the interior of the parcel was covered in feathers. Whatever childish fantasies they might have been cherishing about what the package contained were about to be utterly destroyed when – digging in among the feathers – they uncovered the headless trunk of a woman. Screaming, they ran home to tell their father what they had found. After he had confirmed their story, he immediately called the police.

The following day, two crab hunters were searching the banks of the Hudson at Weehawken, New Jersey, about three miles downriver from where the first package had washed up. They too came across a parcel, one that contained the lower part of a torso. As with the previous parcel there was also a pillow stuffed inside, along with a large rock to weigh everything down. The remains themselves had been wrapped in a newspaper dated 31 August 1913.

Both sections were taken to Volk's Morgue in Hoboken where Dr George W. King examined them. He estimated the woman's age at around thirty, due to the softness of the cartilaginous joints, and suggested that she must have been about 5 feet 4 inches in height, probably weighing approximately 120–130 pounds. He also concluded that the woman had been dismembered by an experienced hand, and that she had also only been in the water for a few days. She had given birth prematurely not long before she died.

Perhaps unexpectedly, it was the rock used to weigh down the second parcel that was to provide a useful lead to investigators. Geologists determined that it was a piece of schist, a greyish-green rock that was rarely found in New Jersey but that was very common in Manhattan. It was irregularly shaped, as if it had been broken off a larger piece by blasting, which, given the massive building programme that was underway in New York at the time, would also be consistent. As a result, after some initial disputes regarding who was responsible for it, the case was finally handed over to the New York Police Department (NYPD).

It was one of New York's finest detectives, Inspector Joseph A. Faurot, who took it on, assisted by detectives first class Frank Cassassa, Richard McKenna and James O'Neil. Faurot was a great believer in forensic science, and had travelled to London in 1906 in order to observe Scotland Yard's use of fingerprinting. Indeed, later that same year, having returned to New York, he arrested a man who had been seen acting suspiciously in the Waldorf-Astoria. The man, who had a British accent and claimed that his name was James Jones, insisted he was only there because he was having an affair with one of the hotel's guests. Not easily convinced, Faurot sent the man's fingerprints to Scotland Yard, where they were matched with those of an infamous hotel thief, Daniel Nolan. Incidentally, this was the first time in American legal history that fingerprints were used to find a suspect guilty.

Faurot began his investigation by looking more closely at the pillows found inside both the packages. On the pillowcase of one he found an embroidered 'A', about an inch high and obviously the work of an amateur. A tag on one of the pillows themselves had a maker's name: the Robinson Roders company of Newark, New Jersey. Faurot visited the company, where he was told that the pillows had been a bit of a disappointment; they had only sold twelve, all to George Sachs, a secondhand furniture dealer. Sachs, when questioned, also said that the pillows had been very slow movers, and that he had sold only two. One of these sales had been to a woman who, when questioned, seemed highly unlikely to be connected to the crime. The other had been delivered, along with some pieces of furniture, to an apartment at 68 Bradhurt Avenue.

The landlord informed Faurot that two weeks earlier he had rented the flat to a man calling himself Hans Schmidt, who said he was acting on behalf of a female relative; it was apparently for her that he had ordered all the furniture. Faurot had the apartment watched for a week but no one went in, or even showed any interest in it. On 9 September, therefore, the police entered the flat by climbing up a fire escape and jimmying a window open. What they found inside was far from pleasant.

In spite of someone's attempts to remove them, dark stains were still visible on the floor and on the green wallpaper. They were obviously blood. There was a trunk, inside of which was a foot-long butcher's knife and a large handsaw. Both had been recently cleaned. In another Faurot discovered several small handkerchiefs, each embroidered with a letter 'A' identical to the one found on one of the pillowcases. There was also a bundle of letters addressed to one Anna Aumuller. Most of them had come from Germany, but three had return addresses in New York. Faurot visited each of these addresses, interviewing the people named in the letters. His final visit was to St Boniface's Church on Forty-Seventh Street and Second Avenue. The pastor there, Father John Braun, remembered Anna Aumuller well – she was a beautiful twenty-one-year-old Austrian immigrant who had worked as a maid in the rectory until she was sacked for misconduct. He also knew the name Hans Schmidt. Schmidt had been a priest at the church but had recently moved on to another, St Joseph's, at 405 West 105th Street. Faurot hastened over there, arriving just before midnight. It was Schmidt who answered the door. When Faurot introduced himself and told him why he was there, he almost collapsed. Once he had recovered, much to Faurot's surprise he made a full confession.

He claimed to have married Aumuller in a bizarre-sounding ceremony that he had carried out himself (for the obvious reason that, as a Catholic priest, he was officially unable to marry). Shortly afterwards, on 2 September, he killed her by slitting her throat while she slept. The only explanation he was able to give for his actions was 'I loved her. Sacrifices should be consummated in blood.' It is perhaps closer to the truth that, after discovering she was pregnant, he murdered her to avoid the matter becoming public. He also admitted to buying both the handsaw and the knife, and when asked why the cuts were so professionally done, he explained that he had been a medical student before being ordained. He said that having dismembered Aumuller's body, he had thrown all the various parts into the river – though no more of them were ever discovered.

When Schmidt's past was investigated it became clear that he had always been troubled. Born in Aschaffenberg, Germany, in the diocese of Mainz, he was ordained there in 1906. He was later arrested for fraud but was then declared insane and released. The local bishop defrocked him, meaning that the papers he took to the US with him were false. In 1909 Schmidt travelled to America, where he presented his papers and was assigned to St John's parish in Louisville, Kentucky. However, after several serious arguments with another priest there, he was moved to St Boniface's in New York City.

Apart from the murder of Aumuller, further investigation revealed Schmidt to have had a second apartment set up as a counterfeiting workshop where, with the assistance of a dentist, Dr Ernest Arthur Muret, he forged $10 bills. Faurot also suspected him of the murder of Alma Kelmer, a nine-year-old schoolgirl whose body was found buried in the basement of St John's Church in Louisville, the church to which he had originally been attached. Her body had been burned, but from the remains authorities suspected that the killer had initially tried to dismember her. The janitor, Joseph Wendling, had been convicted of the crime and sentenced to life imprisonment, but serious doubts persisted about his guilt. It also later transpired that German police in Aschaffenberg wanted to interview Schmidt regarding the murder of a young girl.

Faurot now began to think about preparing for the trial. It was necessary to prove the identity of the remains once and for all. Luckily he managed to persuade a girl called Anna Hirt to take a look at the remains. She was another of the servants at St Boniface's Church and therefore knew Aumuller very well. She explained to Faurot that Aumuller had a brown mark on her chest, and indeed when she was shown the remains she pointed at once to just such a mark. Aumuller's identity was thus established beyond all reasonable doubt and the trial could go ahead.

Schmidt was convicted of the murder on 5 February 1914 and sent to the electric chair two years later, on 18 February 1916. He remains the only Catholic priest in US history to be executed for murder (if indeed he was still truly a priest at the time of the murder).

Another pre-eminent figure in the history of the analysis of human remains is Sir Bernard Henry Spilsbury (1877– 1947), a British pathologist who is considered by many to have been the greatest medical detective of the twentieth century. He was born in Leamington Spa, Warwickshire, and was the eldest of four children. His father was James Spilsbury, a manufacturing chemist. In 1896 he went up to Magdalen College, Oxford, to study natural sciences, and in 1899 entered St Mary's, Paddington, as an exhibition student, specialising in the then novel science of forensic pathology. In October 1905, when the London County Council requested that all general hospitals in its area appoint two qualified pathologists to perform autopsies following sudden deaths, Spilsbury was appointed resident assistant pathologist at St Mary's Hospital.

St Mary's Hospital in Paddington. The hospital was asked to appoint two resident pathologists to handle autopsies and investigations into suspicious deaths. In the course of his work here Bernard Henry Spilsbury solved some of the most horrific murder cases the hospital ever saw.

As a result of his expertise, Spilsbury became involved in the investigation of some of the most notorious crimes of the twentieth century, including that of Dr Hawley Harvey Crippen in 1910, The Brides in the Bath in 1915 and the infamous Brighton Trunk Murder in 1934. However, the case that Spilsbury later confessed was the most challenging he ever encountered is known as the Murder at the Crumbles.

The Crumbles, a shingle beach between Eastbourne and Pevensey Bay, had already been the setting for violence and foul play when, in 1920, two men called Jack Alfred Field and Thomas Gray had killed a young typist called Irene Munro there. Four years later another, more macabre, murder was to follow.

Along the beach were a few orderly cottages, once owned by the local coastguards but now available to holidaymakers for three and a half guineas a week. In April 1924, a man using the name of 'Walter' took up the lease on the cottage known as 'Officer's House' for two months. His real name was Patrick Mahon, and he had hired the cottage as a secluded love nest for him and his mistress, Emily Kaye.

Kaye, a mature, attractive blonde woman of thirty-seven, was a shorthand typist – just like Irene Munro – and arrived in Eastbourne on 7 April. She was pregnant with Mahon's child. She moved into the bungalow expecting it to be the beginning of an exciting new life with Mahon, whom she had met when working at an accountancy firm in London; they had quickly embarked on an affair.

Kaye was fully aware that Mahon was married – to an Irishwoman called Mavourneen – but this did not diminish her attraction to him, particularly as he had led her to believe he was unhappy in matrimony and would soon leave his wife. Kaye was also aware that Mahon was an ex-convict, having been jailed for five years for a bank raid when he was younger. However, she was pregnant, in love, and thrilled at the prospect of a new start with this dark, handsome Irishman. What she was not aware of was Mahon's incessant womanising, and the fact that in addition to bank robbery, he had indulged in fraud at various points in his past.

Mavourneen, on the other hand, was certainly aware of her husband's numerous indiscretions, but seemed willing to stand by her man. Mahon himself, however, had pushed his luck too far: he now faced the reality of having impregnated a woman who expected him to leave his wife – not something he was planning to do.

Indeed, Mahon continued to return to Mavourneen on most

weekdays. He even found time to engage in a new affair, this time with a pretty young woman in Richmond named Ethel Duncan, whom he agreed to take to dinner the following week. All the while, he was concocting a horrifying scheme to deal with Kaye. On 11 April he went to Eastbourne, moving Kaye's trunk to the Officer's House where she was staying. Telling Kaye he was returning to London to arrange a passport application, he in fact went to an ironmonger's in Victoria where he acquired a butcher's knife and a tenon saw. He returned to the Crumbles that same evening and spent the next three nights with Emily. On the evening of Tuesday 15 April, he bludgeoned his lover to death, swept her body into the spare room and locked the door.

Next, in one of the most extraordinary parts of the case, while Kaye's body was still in the spare room slowly decomposing, he invited his new lover, Ethel Duncan, to stay in the cottage over the Easter weekend. She agreed. Mahon now knew he would have to work quickly. He returned to the cottage on Good Friday, prior to Duncan's arrival, and began to dismember Kaye with the knife and saw he had picked up in London. This done, he wrapped up each of the portions of the body and stowed them away in the trunk before leaving this in the spare room once more.

That evening Mahon met Duncan at Eastbourne station and the pair went on to spend an apparently normal weekend together at the cottage. Duncan even saw the trunk after wandering into the spare room. Mahon, in a slight panic, told her it was full of rare books that he was looking after for a friend, before screwing the door shut to prevent further awkward questions. On Easter Monday Duncan returned home, still oblivious to the fact that she had spent the weekend metres from a corpse.

Once she had left, Mahon continued his efforts to dispose of the body. He put the head and several other body parts into a fire. He cut the torso into smaller pieces and boiled it in saucepans to render it down. Finally, he carried some remaining parts of the body to London in a Gladstone bag and dumped them just outside Waterloo Station. It was here that Mahon made his first and only mistake. He left the bag at the luggage office in the station. Shortly afterwards his wife, who knew of his tendency to see other women, found the luggage ticket while searching his suit for clues of infidelity. Her suspicions aroused, she hired a private detective named John Beard to look into the matter further.

On 1 May, she and Beard travelled to Waterloo together and collected the bag. When they opened it they discovered bloodstained clothing, a butcher's knife, and a canvas tennis racket bag bearing the initials EBK inside it. Beard, who was an experienced detective, called the police at once. Mavourneen, still not aware of the serious implications of what had been discovered, was instructed to go home and replace the luggage ticket in Mahon's suit without saying anything about it to him. They then returned the bag to the luggage office and set a trap.

On 2 May, Mahon returned to the station, intending to pick up the bag and return to Eastbourne with it. But the police had been waiting for him. As soon as he had his hands on the bag, two detectives arrested him and took him to Cannon Street Police Station. There they opened the bag and confronted him with the contents. At first he claimed that the blood had come from some meat that he had carried home. However, when he was told that a forensic examination of the bag had shown the blood to be human, he broke down and confessed to the killing, and to attempting to dispose of the body. However, he maintained that Kaye's death was accidental, saying that during a quarrel she had fallen heavily, hit her head on a coal bucket and died. He said that he had panicked, thinking that he would be branded a murderer, and so had decided to hide the body instead.

Two police inspectors were sent to check out the house at the Crumbles. They could tell before they had opened the door that a significant amount of the body must still be inside; even just approaching the house, the stench was overwhelming. Spilsbury was therefore sent for immediately. He described the scenes he found within as 'the most gruesome I have ever encountered'. Inside the trunk already mentioned, he discovered four parcels, each containing various parts of the deceased. There were two large saucepans of boiled human flesh, as well as saucers and other receptacles swimming with greasy human fat. In a hatbox there were thirty-seven different portions of flesh hidden away, while a biscuit tin was found to contain various organs. The carpet was drenched with blood.

It took Spilsbury several days to complete his search of the property. In that time he recovered no fewer than a thousand fragments of calcined bone that he found amid ashes. Each item was carefully catalogued before being removed to his laboratory for scrupulous testing. In the end, every piece of the body was found save the skull and a small portion of one leg. It was Spilsbury who established from her breasts that Kaye had indeed been pregnant at the time of her death, though her uterus, which would have proved the same thing, was missing. However, he could never prove how she died, especially as he was unable to examine her skull. Nevertheless, he was convinced that Mahon's story that it had been an accident was a lie. Quite apart from the grisly treatment of the corpse suggesting a man capable of extreme and disturbing acts, Spilsbury observed that the coal bucket on which Kaye was supposed to have hit her head was completely undamaged, which seemed odd, given the force with which her head would have had to hit it in order to kill her.

Strenuous efforts continued to be made to track down the missing parts of the body, especially the head. The garden was dug up and the beach was searched but nothing further was ever found. However, while remanded in custody, Mahon apparently told another inmate that he had burned the head in the stove during a storm, and that during the procedure it had rolled to face him and its eyes had popped open. In fact, this was a natural effect of the heat and other conditions that the head was under, but Mahon had run screaming from the cottage in terror. Having been given this information, Spilsbury decided to look into whether a head could be totally destroyed by fire. He therefore burned a sheep's head, and discovered that in just four hours it had become a charred remnant, which he was then easily able to smash into dust using a poker. It seemed, therefore, that it was entirely possible that Mahon had successfully disposed of the head in this way.

Mahon was tried for the murder of Emily Kaye on 15 July at Lewes Assizes. He stuck to his story that it had been a tragic accident, after which he had panicked. However, there was considerable evidence against him, perhaps the most damning being that he had bought the saw and the knife he used to dismember the body prior to Kaye's death – a strong indication of premeditation. He even contested this, claiming to have bought them afterwards, on 17 April, but a carbon copy of the receipt clearly showed that he had purchased them on the 12th. It only took the jury forty minutes to find him guilty as charged, and he was executed on 9 September 1924.

One of the most important long-term changes that arose as a result of this case was the creation of the 'murder bag' for use by police. Spilsbury had been shocked to see police officers having to remove rotting flesh and body parts from the scene of the crime using their bare hands. To address this problem, a series of meetings were held between Scotland Yard and Spilsbury, which led to the development of the murder bag, which contained rubber gloves, tweezers, evidence bags, a magnifying glass, compass, ruler and swabs. Such a bag is now an essential part of any major inquiry and may contain various items, depending on the specific department. Common modern additions are: a fibreglass brush, lifting tape, powder, utility knife, scissors, a blood test, a semen test, swabs, alcohol hand spray, scalpels and goggles (see Plate 10).

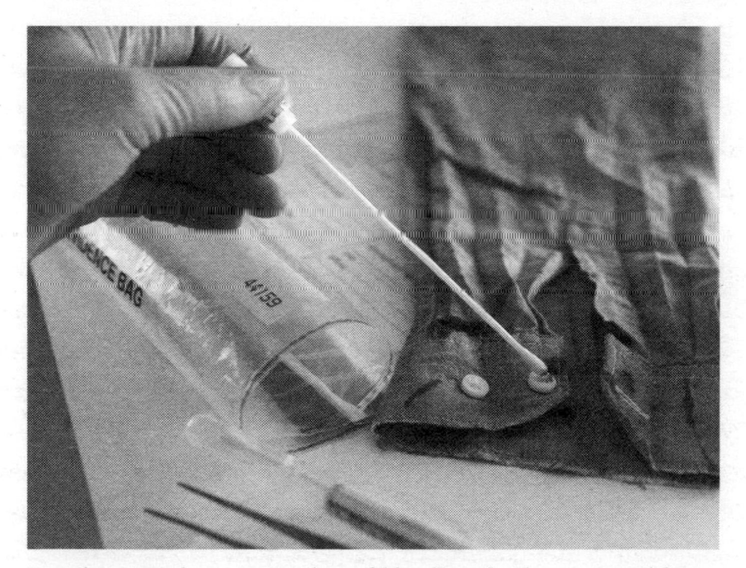

In modern criminal investigations the careful handling of evidence is essential forensic practice. The forensic investigator shown here uses gloves, a swab and a sterile evidence bag to ensure that key materials do not become contaminated in the course of being removed from the crime scene.

Of all the cases I have ever read about, that of Buck Ruxton is perhaps one of the most intriguing and disturbing. It was solved by yet another extraordinary forensic scientist, Professor John Glaister Jr (1892–1971). During the First World War, Glaister served in Palestine with the Royal Army Medical Corps, before returning home to Glasgow in 1919. There, he became an assistant in Glasgow University's Department of Forensic Medicine, where his father (John Glaister Sr, also a prestigious forensic scientist) was Regius Professor. Following this, Glaister Jr spent three years in Cairo as Professor of Forensic Medicine at the University of Egypt, where he had the unique opportunity to examine mummified bodies. He succeeded his father as Regius Professor in 1931. Glaister was in considerable demand as an expert witness, with his most celebrated case undoubtedly being that of Buck Ruxton.

On 19 September 1935, a young woman called Susan Haines Johnson, who was visiting from Edinburgh, decided to take an afternoon walk near the town of Moffat in Dumfriesshire, Scotland. As she was crossing the aptly named Devil's Bridge, over a little stream called Gardenholme Linn, she noticed a bundle of some kind trapped against a rock in the stream. Looking more closely, she realised to her horror that there seemed to be a human arm sticking out of one side of it. She hurried immediately home to the house of her brother Alfred, who called the police.

Inspector Strath and Sergeant Sloane of the Dumfriesshire Constabulary began the investigation. When a search of the immediate area was conducted, several more parcels were found along the banks of the stream, all of which contained human remains. These included an armless torso, a thighbone, legs, pieces of human flesh, and two upper arms wrapped in a woman's blouse and in newspaper. When opened, the latter turned out to be the *Sunday Graphic*, dated 15 September 1935. Two severed heads were also discovered, one of which was wrapped in a child's romper suit.

The following day Glaister arrived at the scene, along with his colleague, Dr Gilbert Millar. Almost immediately they knew that the dismemberment of the bodies had been carried out by someone who knew what they were doing; the dissection was careful and accomplished. A knife had been used – unless a person understands how the human body works, it is almost impossible to cut a body up using a knife rather than a saw. Additionally, the flesh had been peeled from the faces in a clear attempt to conceal the identity of the victims. The fingers had all been sliced off at the terminal joint to prevent fingerprinting being used. Likewise all the teeth had been removed, rendering dental records useless. It later transpired that any marks on the bodies such as birthmarks and scars from operations or injuries had also been carefully removed.

The remains were removed to the anatomy department of the University of Edinburgh and treated to destroy the maggots that had infested them, and to prevent any further decomposition. They were placed in a formalin solution in order to preserve them as much as possible. Glaister and Millar, along with Sydney Smith and James Brash, professors of forensic medicine and anatomy at the university respectively, now began to work on the remains. They were faced with an unenviable task: trying to reconstruct seventy pieces of human body into their original complete forms and then identify them. A macabre jigsaw puzzle indeed.

Their first task was to work out which pieces belonged with each other and to separate them out accordingly. They then began the process of putting them back into a semblance of their original shape. In so doing, they established that one of the bodies was six inches taller than the other, which speeded up the process considerably. They found that they had most of the taller person, but that the trunk of the shorter person was still missing. They also found a large eye, which clearly did not come from either of the victims – Glaister assumed that it had belonged to some kind of animal and had accidentally got mixed in with the bodies of the victims.

Although a lot of questions still remained about who these people were, more body parts kept being discovered as the river was combed, which gave valuable new information. Two hands where the fingertips had not been cut off were discovered. By soaking these in hot water, Glaister was able to get a first-class set of prints off them. At first it had been thought that the smaller remains were those of a male, but as more pieces were recovered, the team eventually had three breasts in their possession, meaning in fact that both bodies must be female.

They now needed to determine the age of the two individuals. Glaister did this through looking at the sutures of the skulls. Sutures are the fibrous joins between the different sections of bone that make up the skull. The process of them sealing up begins in infancy and typically finishes at around the age of forty. In the case of the skull of the smaller person, the sutures remained unclosed; in that of the larger person they were almost completely closed. This meant that the smaller person was certainly under thirty, while the larger person was approximately forty. Further examination of the smaller skull revealed that the victim's wisdom teeth had not yet emerged, which almost certainly meant she was in her early twenties.

The next job was to establish a cause of death for both women. It transpired that the taller woman had sustained five stab wounds to the chest and had a number of broken bones, as well as numerous bruises. The hyoid bone in her neck was broken, suggesting that she had been strangled before the other injuries were inflicted; someone had wanted to make absolutely certain that she was dead. The shorter woman showed signs of having been battered with some kind of blunt instrument, though the swelling of the tongue was also consistent with asphyxia.

While Glaister and his team were working on the remains, the police were out searching for the culprit. The *Sunday Graphic* from 15 September, which the arms had been wrapped in, was an extremely important lead. Not only did it help to establish when the killings had taken place – it also gave a clue as to where. It was a 'slip' edition, meaning it was a special issue for an event of local importance and had been circulated only in that area. This particular edition had been published to celebrate the Morecambe festival and had only been sold there.

Now chance offered a helping hand to the investigation. It so happened that the chief constable of Dumfriesshire read about the disappearance of a woman named Mary Jane Rogerson, a nursemaid at the home of a Dr Buck Ruxton. She had gone missing from Lancaster, which lies close to Morecambe. A telephone call to the chief constable of Lancaster revealed that Ruxton's wife had also gone missing at the same time. This seemed extremely suspicious, and detailed descriptions of both women were quickly forwarded to the Dumfriesshire constabulary.

Buck Ruxton was a Parsi, born in Bombay on 21 March 1899. He obtained his Bachelor of Surgery at the University of Bombay and served with the Indian Medical Service in Baghdad and Basra. His original name was Bukhtyar Rustomji Ratanji Hakim, but he later changed it when he moved to the UK to set up his practice in Lancaster in 1930. Ruxton was respected as a GP, proving popular with his patients. He also regularly displayed generosity, forgoing fees for impoverished patients. His image was that of a family man, and he lived comfortably at 2 Dalton Square with his wife Isabella and their three children.

Jessie Rogerson, who lived in Morecambe and was Mary's stepmother, was brought in by the police to see if any of the clothing that been found with the bodies was familiar to her.

The library at the University of Mumbai (formerly Bombay), where Ruxton studied.

She was distraught, quickly recognising the blouse as one belonging to her stepdaughter, and pointing out a repair that she herself had previously made to it. The child's rompers which had been used to wrap one of the heads were identified by a Mrs Holmes of Grange-over-Sands as a pair that had been given to the Ruxtons some time before. Jessie Rogerson had suggested her as a lead because Mary had holidayed with her that same year and Mrs Holmes had given her the rompers for her children. The unpleasant fact of the matter was that the police now had very good reason to link Buck Ruxton to the case. Being a medical man, he also had the kind of practical anatomical knowledge that the killer must have had to dismember the bodies. The police acted quickly and arrested him.

The last time Isabella had been seen alive was on Saturday 14 September, when she and her sisters had gone to Blackpool, spending time there enjoying the lights before returning home that night. It transpired that on the following Monday, Ruxton had called his cleaner and said that her services would not be required since his wife had gone on holiday to Edinburgh. He did though, rather bizarrely, invite a Mrs Hampshire over, asking if she could help him clean up the house to get it ready for the decorators who were due the following week. She would later testify that she helped him to dispose of bloodstained carpets and clothing by burning them outside in the garden. Quite why she did not consider this suspicious at the time is unclear – perhaps she naively thought blood must be an occupational hazard of being a doctor. Other witnesses also gave evidence that fires were seen burning behind the house for several days. When the police searched the property, human flesh was found in the waste pipes and drains leading from the bath, and bloodstains were discovered on the carpet on the stairs, and on the bathroom walls and floor. When the charred remains outside were searched, several pieces of cloth were recovered which were identified as having belonged to Mary Rogerson.

It seemed certain that the police had their man, but they still needed to positively identify the bodies in their possession as being Mary and Isabella. They were able to do so through several ingenious forensic techniques. Firstly, fingerprints that they had managed to take from one of the bodies were found to match prints they were able to lift from several items in the house that Mary touched on a regular basis.

Now came a first in the history of forensics. The team managed to get hold of photographs of the two women, a studio shot of Isabella and two poorer-quality images of Mary. The skulls were cleaned of any remaining tissue and then photographed from various angles to match the angles shown in the photographs as closely as possible. When these new photographs were blown up to the same size and superimposed over the original photographs of Mary and Isabella, they matched perfectly (see Plate 11).

And finally, in another ingenious first, Dr Alexander Mearns of the Institute of Hygiene at the University of Glasgow was able to establish, through observing the life cycle of the maggots that infested the remains, that the two victims had been killed at approximately the same time that Isabella and Mary had last been seen alive. Never before had entomology been applied forensically like this.

With this weight of evidence against him, it is perhaps unsurprising that Ruxton was found guilty of murder, though he protested his innocence to the last. He was executed at Strangeways Prison on 12 May 1936. It is believed his motive for the murder must have been jealousy, and the belief that Isabella was being unfaithful to him. The two were known to have a tempestuous relationship and the police had been called out to the property as a result of their arguments on more than one occasion. Poor Mary Rogerson was probably just in the wrong place at the wrong time and witnessed something that she should not have. Ruxton had used his expert knowledge to good effect in trying to conceal the identities of his victims through systematic mutilation, but this was not enough to eradicate all the information that the remains were able to impart.

While the natural impulse of any person when confronted with a dead body – particularly one that has been horribly disfigured or dismembered – is to recoil, when a violent crime is being scrutinised, a corpse often represents the focal point of the investigation. When a body can provide so many different forms of valuable evidence that may bring a killer to justice, a forensic scientist cannot afford to be squeamish.

6

Poisons

A poison in a small dose is a medicine, and a medicine in a large dose is a poison.

Alfred Swaine Taylor, English toxicologist (1806–1880)

Poisoning was described by the Jacobean writer John Fletcher as 'the coward's weapon' – a stealthy way of killing that leaves no mark of violence on the body and that might even be mistaken for illness. Particularly in the past, it is often associated with repressed and marginalised members of society – those who did not have recourse to other methods. For this reason it was also frequently associated with women; a wife might not have been able to physically overpower her husband, but poison gave her a less direct way to end his life. From a forensic perspective, poisons present their own problems and challenges, which we will explore in this chapter. First, however, it is perhaps worth taking a brief look at poisoning throughout history, and tracing the gradual acquisition of knowledge about the practice.

Sources show that ancient civilisations were aware of a

variety of poisons and their effects. There is evidence that as far back as 4500 BC the Sumerians had knowledge of a number of poisons. A Sumerian tablet discovered in 1850 documents the use of poisons as a clandestine way to rid yourself of an enemy. And in Egypt in 3000 BC the first pharaoh, Menes, is reputed to have conducted research into many varieties of poisonous plants and their properties. The ancient Egyptians also possessed knowledge of how to create and refine particular poisons, such as knowing how to extract cyanide from peach kernels. An ancient Egyptian text that was discovered in 1872 contains a list of poisons and their antidotes. Later, in ancient Greece, poison was employed in state executions, where the condemned would be made to drink a cup of hemlock. Famously, this was how the philosopher Socrates was put to death in 399 BC.

Ancient Rome was a hotbed of political intrigue and power games, where seeking the permanent removal of a rival was not uncommon. By 82 BC, poisoning had become such a scourge in the empire that the dictator and constitutional reformer Lucius Cornelius Sulla found it necessary to issue the world's first law against poisoning, the Lex Cornelia. Despite this, poisonings continued to increase, peaking in the first century AD, during the reign of the Julio-Claudian emperors. The Roman historian Tacitus writes about an infamous poisoner by the name of Locusta, who reportedly killed the emperor Claudius with a dish of poisoned mushrooms in AD 54, having been hired to do so by his fourth wife, Agrippina the Younger. The following year she was convicted for a completely unconnected poisoning, but was rescued from execution by the emperor Nero in return for her help in poisoning Claudius's young son Britannicus.

A bust of Lucius Cornelius Sulla, who issued Europe's first known law against poisoning.

And in India under the Mauryan Empire, *visha kanyas* ('poison damsels') were supposedly used as assassins. These Mata Haris of the ancient world would flirt their way into the trust of their victim, only to mix poison in his food or drink. It was also claimed that they saturated their bodies with gradual doses of poison (to which they themselves became accustomed) and that as a result men died after licking their naked bodies. While these salacious tales are unlikely to be wholly true, they nevertheless demonstrate that poisoning was a cultural phenomenon.

This naturally meant that people also began to concern themselves with how to deal with it. Physicians started writing the first forensic works on how to detect poisoners, such as *On Poisons*, a text attributed to the Indian scholar and royal adviser Chanakya, who lived 350–283 BC. In the second century BC the Greek Nicander of Colophon wrote *Theriaca* and *Alexipharmaca*, the two oldest extant works on the subject of poisons, while another Greek, Dioscorides (AD 40–90), classified poisons and differentiated their origins in his medical treatise *Materia Medica*, which for fifteen centuries was the authoritative textbook on the subject of poisons.

One of the most significant – and in many ways regrettable – discoveries in the history of poisoning occurred much later, in the eighth century. Jabir ibn Hayyan, sometimes known as Geber, was a prominent chemist, alchemist, astronomer and general polymath who was born in AD 722 in the Persian city of Tus (modern-day Iran). One of the many fields in which he is credited with making advancements is the alchemical discipline of distillation and crystallisation. The understanding of these everyday chemical processes helped lay the foundations for the modern-day study of chemistry, but among the substances that Geber succeeded in crystallising was arsenic. Transformed into a colourless, odourless and tasteless powder, it became one of the deadliest of all poisons, impossible to detect until at least ten centuries later. It would later acquire the nickname 'inheritance powder', because of its supposed application by impatient heirs.

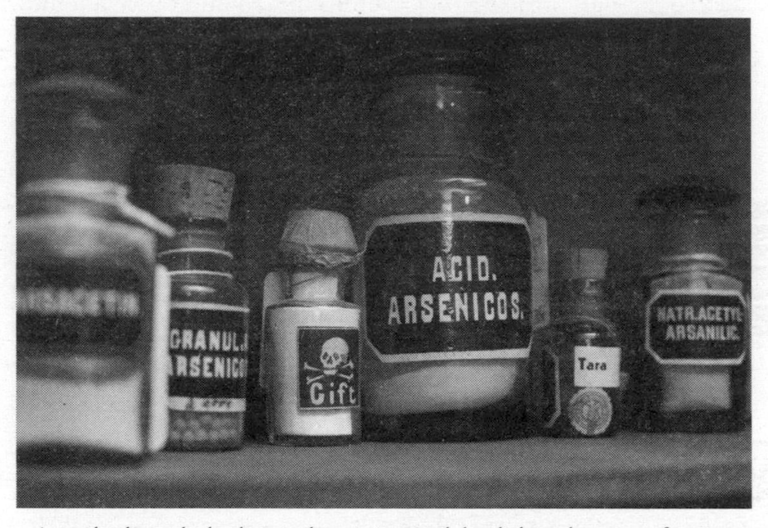

As a colourless and odourless powder, arsenic proved the ideal murder weapon for many centuries – and the weapon of choice for those wanting to rid themselves of a troublesome spouse or relative.

As new poisons were developed, the interest in their use for criminal purposes grew – after all, they were an extremely convenient way of ridding yourself of an enemy whilst avoiding detection. Poison became a growth industry, with people soon setting up businesses selling books on the subject and even supplying poisons to the public. Some engines of state fully embraced poisoning as a method of removing problematic individuals – the Venetian Council of Ten (1310–1797), for example, was notorious for employing the practice. In fact poisoning really took hold in Italy – in the sixteenth and seventeenth century, schools of poisoners began to be set up there; secret societies that taught the poisoner's art. A publication called *Magia Naturalis* (1589) by Giovanni Battista Della Porta served as a textbook for poisoners, mentioning in particular how to lace wine with a preparation know as *veninum lupinum*, a combination of aconite, *Taxus baccata*, caustic lime, arsenic, bitter almonds, powdered glass and honey, shaped into pills.

One of the most notorious women to be associated with poisoning in Italy was a Neapolitan by the name of Toffana di Adamo. She created an arsenic-infused solution called *Acqua Toffana*, which she marketed as a ladies' cosmetic, claiming it was a 'miraculous substance oozing from the tomb of St Nicholas di Bari'. In fact, it was famous among widows for more sinister purposes than beautification. The authorities soon caught on to the scheme as a wave of husbands began suddenly to die. Toffana was caught and admitted to causing the deaths of almost 600 husbands through selling *Acqua Toffana.* She was strangled to death in a Naples prison in 1709.

This was not the only instance of women using poison for murder in seventeenth-century Italy. During the 1650s there was a noted increase in the number of young, rich widows in the larger cities of Europe. Although some of them even confessed to their priests that they had murdered their husbands by poisoning them, the priests were bound by the seal of the confessional. However, the sheer number of these confessions began to alarm them. In 1659 they appealed to Pope Alexander VII for guidance. The pope took the matter very seriously and instigated an independent investigation. What his spies discovered surprised and shocked him. A group of young wives, some from among

Rome's first families, were meeting regularly at the house of Hieronyma Spara, a well-known witch and fortune-teller. She was training these women in the art of poisoning. Papal police arrested La Spara and she and several other women were hanged. A further thirty young wives were whipped through the streets.

Another famous case of poisoning occurred in late seventeenth-century Paris. It bears many similarities to the crimes of Toffana di Adamo and Hieronyma Spara. Two midwives, Mesdames Lavoisin and Lavigoreux, were arrested in late 1679 for teaching, instigating and supplying poison for hundreds of murders all over France. Using their occupation as midwives – as well as a sideline they had in fortune-telling – the two women gained access to people from every social class. On their arrest, a book was found belonging to Lavoisin that contained lists of everyone who had come to her to buy poison, including high-ranking individuals such as Marshal de Luxemborg and Duchess De Bouillon. Their main customers were wives trying to get rid of their husbands and men trying to eliminate relatives so they could inherit their estates. The two were tried, found guilty, had their hands cut off and were burnt alive in February 1680. In spite of the dire repercussions for those caught, the craze for such schools of poisoning lasted well into the eighteenth century.

Despite the frequency with which poison was used as a pitiless and covert way to kill, it was a long time before forensic techniques were used to catch the offenders. But finally, in 1751, a toxicological report was used in the murder trial of Mary Blandy, a woman accused of poisoning her father with arsenic.

Mary was still unmarried at the age of twenty-six, which for the times was very late. It seemed puzzling that she should not have a husband, since she was by all accounts sweet, attractive and charming, and in addition would bring with her a dowry of £1,000. This ought to have made her a pleasing prospect for any man. However, her father, Francis Blandy, a prominent lawyer in Henley-on-Thames, had ambitions for her. He quickly saw off any suitor he felt was either not rich enough or not of high enough social standing. This seemed to be all of them.

Eventually, though, Mary made the acquaintance of a Scottish army captain, the Hon. William Henry Cranstoun. He wasn't a handsome man, having a pockmarked face and cross-eyes, and despite being the son of a Scottish peer was short of money. However, Mary was smitten with him. There was, admittedly, the small matter of him already being married, but not wanting to let a little thing like this stand in the way, he declined to mention it. This arrangement seemed to suit everyone: Francis Blandy was delighted at the thought of his daughter marrying into the aristocracy, Mary was delighted at being with the man she loved and Cranstoun was delighted at having found a comfortable and prosperous billet.

In due course he decided that it was time to sort out the matter of his wife. He wrote to her, requesting that she kindly deny that they had ever been married and say instead that she had only ever been his mistress. Unsurprisingly, Mrs Cranstoun took exception to this, and then took the captain to court. The ensuing publicity around the case exposed Cranstoun's real situation to the Blandys – he was a fortuneless army captain who simply happened to have a title. Francis Blandy was outraged and forbade Mary from ever seeing him again.

Mary, however, would not be put off so easily. The couple began to meet in secret, aided and abetted by Mary's mother (particularly after Cranstoun loaned her £40 to pay off a debt in London).

Then, in 1749, Mary's mother died of a sudden illness. Cranstoun himself was now in debt and pressing Mary to pay back the £40 her mother had borrowed. In the end she was forced to take out a loan herself in order to do so. All in all, things were not going well for the couple. Cranstoun began to consider how much better things would be if Mary were to come into possession of the £10,000 inheritance which she was set to receive when her father passed away. He began to concoct an unpleasant scheme. Perhaps, he suggested to Mary, some kind of potion or remedy might improve her father's disposition so that he would be less hostile to their union. He explained that he even knew of a herbalist who could prepare such a mixture – all they would have to do was lace his food and drink with it. Mary did exactly as he asked and it was not long before Francis Blandy became violently ill, suffering stomach pains and acute nausea. He lost weight rapidly.

The family maid, a girl called Susan Gunnel, gradually became suspicious of the circumstances surrounding his failing health. She tried some of the food that Mary had prepared for him and soon started to feel unwell herself. This seemed to confirm her suspicions and she decided to examine the pan in which the food had been made. When she did so, she discovered a gritty white powder. This she scraped out onto a piece of paper and then took to the local apothecary for his opinion. He, however, had no way of analysing the powder to determine what it was. Nevertheless, Susan returned home and warned Francis Blandy that his daughter was poisoning him. He promptly called Mary to his bedside and asked her whether she was indeed tampering with his food, at which she went white and fled from the room.

In spite of this behaviour, for some reason he allowed Mary to continue to make his meals. As his condition worsened,

Mary called the doctor who, having previously been shown the powder by Susan, told Mary that if her father died she would be accused of murder. Mary quickly disposed of Cranstoun's love letters and the remaining powder. A few days before her father finally succumbed, he asked to speak to Mary. Entering the bedroom, she threw herself on her knees and begged him not to curse her. He placed his hand on her head and told her that he blessed her, and that he hoped God would forgive her.

On 14 August 1751, Francis Blandy finally died. Knowing that she was already under suspicion, Mary offered the footman £500 if he would help her run away to France. He refused and she was forced to flee on her own. It wasn't long, however, before the circumstances of Blandy's death became common knowledge. A hue and cry went up and, despite her best efforts, Mary was captured and brought back to the village. Cranstoun also ran for France when he heard of the situation. He made it, but died there in poverty a few months later.

How we interpret Mary's actions depends on whether we believe she knew that the powder was poisonous all along. If she didn't know, then it is reasonable to think she fled the room when her father asked if she was poisoning him because she feared he would be angry at being administered mood-lightening powder without permission. When told by the doctor that the powder was harmful, she immediately disposed of it. If she was administering the poison in full knowledge, then it is clear she fled from guilt, continued to give her father the powder until he was dying and then destroyed the evidence. However, it is perhaps harder for us to believe that she was totally ignorant of what was happening.

While Mary was awaiting trial she learned that, far from her father leaving £10,000 behind, he had left less than £4,000. This was the real reason that he had not wanted her to marry; he couldn't afford the £1,000 dowry. Since Cranstoun had been interested in Mary on account of the supposed £10,000 she stood to inherit, the sad fact is that if Francis Blandy had been more honest about his situation and explained his predicament to Mary, tragedy might well have been averted. He died for a nonexistent fortune. Perhaps that is why he found it so easy to forgive her.

Mary was tried for murder at Oxford Assizes on 3 March 1752. The trial took only a single day. As well as the testimony of Susan Gunnel regarding the white powder she had found at the bottom of the cooking pot, a cook gave evidence that she had seen Mary throw the letters and white powder onto the kitchen fire. A postmortem had been conducted on Francis Blandy, and although his organs could not be definitively tested for arsenic (as such a test had not yet been devised), the well-preserved state of them led several doctors to suggest that arsenic poisoning was a possible cause of death. An examination of the powder that the cook saved from the fire confirmed that it was indeed arsenic. Admittedly their method for establishing this was somewhat rudimentary: they applied a red-hot poker to the sample and smelled the resulting vapour. Still, they were convinced. Mary was found guilty and hanged on 6 April 1752. Dressed all in black with her hands tied behind her back with a black ribbon, she asked the hangman not to hang her too high for the sake of decency.

Even though medical testimony helped convict her, the reason that Mary was caught was largely because she was not careful enough to conceal what she was doing. Had the evidence of the powder not been discovered by the servants, it is unlikely that anyone could have proved that arsenic had been used to kill Francis Blandy – as we have seen, there were no reliable scientific techniques for detecting the presence of the poison. It was a German-Swedish chemist called Carl Wilhelm Scheele (1742–1786) who changed that.

Scheele was already well known in scientific circles, having discovered seven different acids (in actual fact he discovered many more than this, but always seemed to be beaten to publication by someone else, hence his nickname, 'Hard Luck Scheele'). In 1775 he made the further discovery that it was possible to make an acid by heating arsenic trioxide, 'white arsenic' powder, in a solution of nitric acid and zinc. This created a dangerous gas that smelled of garlic – arseniuretted hydrogen, or arsine. This discovery meant that Scheele could now perform a postmortem test to determine if someone's stomach contained arsenic trioxide.

The German physician Samuel Hahnemann (1755–1843) is best known for developing the system of alternative medicine known as homeopathy. However, in 1785 he also found a method of detecting the presence of arsenic that differed from that developed by Scheele. Hahnemann discovered that when sulphuretted hydrogen gas (the gas that smells like rotten eggs) is bubbled through an acidified arsenic solution, the result is a yellow deposit: arsenic trisulphide. In Hahnemann's test a sample that was suspected of containing arsenic could therefore be tested in two simple steps. 1. Dissolve the sample in nitric acid. 2. Bubble sulphuretted hydrogen through the solution. If a yellow sulphide appears, arsenic is present in the sample.

And in 1787, Johann Daniel Metzger (1739–1805), a professor of medicine at the University of Berlin, discovered an even easier way to confirm the presence of arsenic. He found

Samuel Hahnemann, who invented a simple test for the detection of arsenic and thereby paved the way for the arrest of many poisoners.

that when a suspect material was heated along with charcoal, arsenious oxide would vaporise if it were present, and that it would leave a shiny black deposit on a porcelain plate (known as an 'arsenic mirror') if you held it above the heated mixture.

However, while all these discoveries were obviously important steps forward, they were only really theoretically useful. As yet there was still no way to apply these tests in a practical, forensic context. This problem was solved in 1806 by Dr Valentine Rose of the Berlin Medical Faculty. He cut up the stomach of a victim who had allegedly been poisoned with arsenic and boiled it in water. He then filtered the resulting liquid and treated it with nitric acid. This would have the simultaneous effect of ridding the mixture of any remaining traces of flesh and converting any acid in the mixture into arsenious acid. Following this, he used potassium carbonate and calcium oxide, which would turn arsenious acid to arsenic trioxide. He was then able to perform Metzger's test to confirm the presence of arsenic – although it could have easily been detected through any one of the various means we have already mentioned.

We now come to one of the great unsung heroes of crime detection, the chemist James Marsh (1794–1846), who eventually held the post of Ordnance Chemist at the Royal Arsenal, Woolwich, during the 1830s. In 1829 he had worked as Michael Faraday's assistant and had apparently shown a great deal of promise as a scientist. It was a few years after this, in 1832, that Marsh was called to test a powder found in the organs of George Bodle, an eighty-year-old man with a vast fortune of £20,000 (about £2 million today). The prosecution believed this powder was responsible for his death.

Bodle was a farmer from Plumstead near London and was

ordinarily a vigorous and healthy man. However, one day he was suddenly taken ill after drinking his morning coffee. He began vomiting, suffered stomach cramps and subsequently died. The local justice of the peace, Mr Slace, began to investigate the matter. He soon found out that Bodle was not a popular man with his family; that he was dictatorial and prone to fits of violence. It was also noted that nobody in the family seemed upset at his death. There were rumours circulating that Bodle's grandson John had wanted him dead, and the sooner the better. Given these suspicious circumstances, Slace asked Marsh to test the contents of George Bodle's stomach, and the coffee, in order to establish whether or not arsenic could have been the cause of death. Marsh used Scheele's test, which quickly revealed the presence of arsenic in the coffee. Likewise, when the stomach contents were tested, a yellow sulphide precipitate confirmed that arsenic was present. A local chemist also testified that he had sold John Bodle arsenic trioxide, while a maid who worked on the farm reported that John Bodle had said he wanted his grandfather dead so he could inherit his money. The results from Marsh's tests combined with the witnesses' statements seemed to confirm John Bodle's guilt. It was an open and shut case.

However, at John Bodle's trial in Maidstone in December 1832, he was surprisingly found not guilty. Part of the reason for this was that the samples of yellow arsenic trisulphide that Marsh had recovered from the coffee and stomach contents had deteriorated by the time of the trial. The jury therefore had to acquit on the grounds of reasonable doubt, since there was no incontrovertible proof of the presence of arsenic. Many years later John Bodle, who in the intervening time had been deported to the colonies for fraud, finally admitted to murdering his grandfather, though too late for anything to be done about it.

At the time, Marsh was stung by his failure to secure a conviction, and resolved that he would pick up where Scheele had left off. He wanted to create a test that could not only infallibly detect the presence of arsenic, but which would also be able to be understood by a lay jury. The method he eventually developed involved adding the sample matter to a solution of hydrochloric acid and zinc, which would create arsine gas if arsenic was present, as well as hydrogen gas generated from combining the zinc and acid. If he trapped this gas, directed it through a tube, and then ignited it, a silvery black stain would form on a porcelain plate held in front of the tube if arsine gas was present – metallic arsenic. The test did prove to have its drawbacks, though; if antimony (another poisonous substance) was present, it would also form a black deposit under this test. However antimony, unlike arsenic, dissolves in sodium hypochlorite, so there was a way of differentiating the two if necessary.

Marsh's test proved to be so sensitive that quantities of arsenic as small as one fifth of a milligram could be detected. He first published the details of it in *The Edinburgh Philosophical Journal* in 1836. Marsh gave us the first really practical and reliable test for the presence of arsenic. He died in 1846 aged fifty-two, leaving his wife and children destitute. A sad end to such a bright and influential career.

In 1841 a German chemist named Hugo Reinsch (1809–1884) published a new arsenic test, which required far less skill than the Marsh test. Because of its simplicity, it promised to provide far more accurate results when carried out by less experienced chemists. For the Reinsch test, a sample of the suspected liquid was mixed with hydrochloric acid. Then a polished copper foil strip was placed into the mixture. If any arsenic was present in the sample, it would react with the hydrochloric acid and leave a grey stain on the copper foil. But although simpler and quicker to perform, this method soon proved to have its own drawbacks, as demonstrated in the trial of Thomas Smethurst.

On 2 May 1859, Smethurst, a retired surgeon in his late forties, was arrested for attempted poisoning. A few years earlier he had moved into a boarding house in Bayswater, London. It was there that he met Isabella Bankes, a fellow lodger. Isabella was a wealthy, independent woman, also in her forties. It did not take long for Smethurst to begin a passionate affair with her. He soon left his wife to begin a new life with Isabella in Richmond, and on 9 December 1858, Smethurst and Isabella were married, despite Smethurst still being bound by his previous vows.

In March 1859, only a few months after their marriage, Isabella became violently ill, exhibiting symptoms such as high fever, vomiting and diarrhoea. She was treated by several doctors, and when her condition did not improve, they became suspicious and sent samples of her bodily excretions to be examined. These were tested by Alfred Swaine Taylor, the widely respected English toxicologist. He found what he believed to be a metallic poison in one sample, and Smethurst was arrested on suspicion of poisoning. Unfortunately it was too late to help Isabella and she died soon after his arrest.

After her death, Taylor carried out further tests, including on medicine bottles found in the rooms she shared with Smethurst. One medicine bottle containing chlorate of potash was also found to contain traces of arsenic. Taylor deduced that Smethurst had poisoned the medicine so he could administer it to Isabella undetected. Smethurst was immediately charged with murder. However, Isabella's postmortem failed to find any trace of arsenic in her body, and nor did further tests on bodily fluids that had been collected when she was alive.

Confused, Taylor tried to think of another explanation. It was known that copper could sometimes contain arsenic impurities. Taylor suddenly realised that the copper he had used during the Reinsch test was not pure enough, and that arsenic impurities in it had reacted with the hydrochloric acid used in the test to create the telltale grey stain. Smethurst's trial was already under way, and despite Taylor alerting the court and the jury to his mistake, it continued, as the magistrate believed the 'metallic poison' found in the original sample from Isabella was enough evidence to prove Smethurst's guilt. Smethurst was found guilty of murder. However, after vociferous protest from both the public and medical professionals – including petitions to the Home Office – his sentence was lifted. Commentators on the case highlighted evidence that seemed to have been ignored during the trial. Isabella was prone to fits of illness throughout her life, including extended periods of vomiting. Not only that, but the medicine she received from doctors in fact contained mercury, which might well explain the metallic poison that was believed to be present in her fluid samples. It seems likely that she simply died of natural causes, a consequence of her chronic poor health.

This case highlights the fact that, despite the relative ease with which the Reinsch test could be performed, an expert knowledge of chemistry was still needed to ensure an accurate result. Both the Marsh and the Reinsch test were valid and effective methods of testing for arsenic, but only when the procedure was carried out correctly.

* * *

Another significant case in the history of forensic toxicology is that of the Frenchwoman Marie-Fortunée Lafarge, who was convicted of poisoning her husband in 1840. The case became a cause célèbre, as it was one of the earliest trials to be reported on daily by newspapers (see Plate 12), and because she was the first person convicted largely on direct forensic toxicological evidence – as opposed to the more indirect evidence that Mary Blandy was convicted on.

Marie-Fortunée Lafarge (née Chapelle) was born in Paris in 1816, the daughter of an army officer. She was said to be, through her grandmother's line, a direct descendant of King Louis XIII. She lost both her parents when she was still young and was eventually adopted by her maternal aunt, though she was already eighteen. The two did not get along and although Marie was educated and treated well, she was always made to feel like *une cousine pauvre* – 'a poor cousin'.

When she reached the ripe old age of twenty-three and was still unmarried, her aunt took it upon herself to find Marie a husband. However, she failed to inform Marie of this little project. She engaged the services of a marriage broker, who eventually came up with a candidate who fitted the bill. His name was Charles Lafarge, and he was the son of a justice of the peace. He was twenty-eight years old and apparently a giant of a man, coarse and uncouth. His family had been left penniless by a succession of poor business deals and his father saw the marriage as a way for them to get their hands on some more cash. Obviously he was not keen for Marie's family to learn of this ulterior motive, so he concealed the reality of the family finances, claiming that they were wealthy in both land and business.

Since all seemed satisfactory, a meeting was arranged between

Charles and Marie. Unfortunately she was unimpressed by him. Nevertheless she eventually agreed to marry him, since she was under the impression that he was a rich man. Their engagement was announced and on 10 August 1839 they became husband and wife, then left Paris to begin life at the Lafarge estate.

Marie was horrified when she arrived; it was immediately clear that she and her family had been tricked. The house, which was contained within the ruins of a former monastery, was in a very poor state of repair, damp and rat-infested. There was no money, just a mountain of debt. Marie considered her new family little better than vulgar peasants. She locked herself in her room and wrote a letter to her new husband begging him to release her from the marriage and threatening to kill herself if he did not. He refused, but promised not to assert his marital privileges until the estate was restored. Marie also insisted that they each make a will leaving everything to the other. Charles did this but then promptly changed it to leave everything to his mother.

Then, while Charles was away in Paris trying to raise money for a new business venture, Marie unexpectedly sent him a Christmas cake along with – even more unexpectedly – a love letter. After eating the cake he became violently ill. He did not contact a doctor, however, and assumed that the cake must have been contaminated in some way while it was being transported. He returned to the estate, still feeling very unwell. Marie insisted that he take to his bed and that he should allow her to nurse him. She prepared all his food, and the illness immediately flared up again. The family doctor judged that it seemed a little like cholera.

Charles continued to experience a variety of symptoms including cramps, dehydration and nausea. He became so ill that the family decided that he should be watched twenty-four hours a day. A young cousin called Emma Pontier and a family friend called Anna Brun were chosen to help look after him. All the while, Marie continued to treat him with various medicaments, including gum Arabic which she said she swore by. In spite of this apparent care, however, Charles continued to deteriorate. He was prescribed eggnog to keep his strength up. It was Anna Brun who noticed Marie taking a white powder from a malachite box she owned and stirring it into the eggnog. When she asked what it was, Marie told her it was 'orange blossom sugar', which she was adding to the drink to sweeten it. This reply failed to satisfy Anna, and when she noticed a few white flakes still floating on the surface, she began collecting samples of the food that Marie prepared for Charles as evidence.

After two weeks spent suffering in extreme pain, Charles passed away. Marie seemed totally unfazed by her husband's death, and remained calm even when informed that the police were being sent for. Two days later, a justice of the peace named Moran arrived from Brive. He immediately took possession of the soup and eggnog that Anna Brun had kept. Moran had heard of a new test that pathologists in Paris were using to detect the presence of arsenic. This was, in fact, the Marsh test. He asked the doctors who had treated Charles if they knew of it. Not wishing to seem foolish or uninformed, they said that they did – in fact they had never even heard of the method, let alone carried out such a test themselves.

The doctors performing the autopsy had only removed the stomach before burial. They performed an old-fashioned and unreliable test that involved heating the sample to try to determine the presence of arsenic. When they did so, it gave off a strong smell and formed a yellow precipitate, but the test was performed so incompetently that the test tube actually exploded. Still, they eventually concluded that there was a high concentration of arsenic in the body of Charles Lafarge.

Moran then turned his attentions to Emma Pontier. When questioned she admitted to hiding Marie Lafarge's small malachite box. When the box was recovered it was handed over at once to a Dr Lespinasse for examination. He found that the substance it contained was indeed arsenic. Moran also learnt that before Charles had received the cake that made him ill, Marie had procured arsenic from a local chemist, claiming it was to treat a rat infestation. With this development, Moran felt compelled to act – he arrested and charged Marie with murder. She was incarcerated in the jail at Brive.

The trial took place in Tulle on 3 September 1840. A young lawyer called Charles Lachaud was appointed to defend Marie, assisted by several others. One of these, a man called Maître Paillet, was acquainted with the toxicologist Mathieu Orfila, who was, among other things, an acknowledged expert on the Marsh test. The strongest evidence against Marie was the test carried out by the doctors at Brive, which had concluded that there was arsenic present in the corpse. When Paillet wrote to him explaining how the results had been obtained, Orfila was furious at the outdated methods employed. He sent an affidavit to the court stating that the tests were conducted so ignorantly that they meant nothing.

The doctors' report stated that when they performed their test, it resulted in a yellow precipitate that they assumed to be arsenic. However, Orfila argued that as the test had been carried out incorrectly, the yellow precipitate could actually have been caused by many other substances and could not be considered conclusive evidence. When Orfila's statement was read in court, it destroyed the evidence of the doctors from Brive. In an unexpected move the prosecution then insisted that the stomach contents be subjected to the Marsh test as Orfila suggested.

Three chemists from Limoges were found who, despite their inexperience, attempted to carry out the test. However, when they did so, they found no traces of arsenic in the stomach sample. This was not a welcome development for the prosecution – they were keenly aware that the balance of evidence was tipping, and not in their favour. However, they had a last ace up their sleeve: they asked that the food that Anna Brun had hidden away should be similarly examined. The defence were by now feeling confident in their victory and so acquiesced. Unfortunately for Marie, this time the tests did find arsenic, and in prodigious quantities. This was a puzzling development – if there was poison in the food, then why was there no poison in the body? Finally it was decided that Orfila should be called personally.

A week later, Orfila arrived, and conducted the Marsh test on the contents of Lafarge's stomach himself, insisting that the three chemists watch his methods carefully. By the afternoon of the following day, Orfila had his answer: the previous tests had been incorrectly performed. There was arsenic in the body of Charles Lafarge, although admittedly in small quantities.

The defence tried in vain to discredit these new results, but to no avail. On 19 September 1840, Marie Lafarge was found guilty of murder and sentenced to life imprisonment with hard labour. She was transported to Montpellier to serve her sentence. King Louis-Philippe later commuted the hard labour (but not the life sentence). In 1841, while still in prison, Marie wrote her *Mémoires*, which were published later that year. In them she completely denied any wrongdoing. In 1852, Napoleon III finally released her as an act of mercy, since she was suffering from tuberculosis. She died on 7 November that year, still proclaiming her innocence.

The case highlights just how important accurate toxicological testing could be. Had Orfila not become involved, the inaccurate results of the previous tests could well have led to a completely different outcome in the trial. It also led to King Louis-Philippe issuing a decree forbidding apothecaries to sell arsenic or any other poison to anyone not already known to them. Anyone buying a substance that could be used as a poison was also required to sign a register, a 'poison book'. The rest of Europe soon adopted similar measures.

Although arsenic was undoubtedly pre-eminent among poisons, it was by no means the only substance used during this period. In fact, during the early years of the nineteenth century, new poisonous substances were being developed at an alarming rate: strychnine in 1818 and chloroform in 1831, for example. Even years later, in 1847, Mathieu Orfila confessed that he thought it possible that vegetable poisons would remain undetectable in the body, and that there was nothing that he or anyone else could do about this problem. This is because the basis of vegetable poisons are alkaloids such as morphine, strychnine, nicotine. These work in the nervous system and therefore leave no trace in the body that scientists in the early nineteenth century were able to detect. Fortunately for forensic science, he was wrong on this count. A few years later a Belgian chemist called Jean Servais Stas (1813–1891) came up with a solution while working at the École Royale Militaire in Brussels.

Gustave Fougnies had no title, yet he was a wealthy man, having recently inherited the considerable fortune of his father, who had been a greengrocer. On 20 November 1850, he collapsed and died on the dining-room floor of the Château Bitremont where his sister Lydie lived in Belgium. This was, in itself, perhaps not entirely remarkable: he had been weak since birth, and had even had his leg amputated as a result of his poor constitution – something that had only made his condition worse. What was remarkable was the apparent indifference of his sister to his death – if anything, she seemed pleased. However, once you dig into the family situation in a little more depth, the reason for this quickly becomes apparent.

Lydie was a countess, having married Count Hippolyte de Bocarmé, who was descended from the noble Belgian family of Visart de Bocarmé. Despite his title, Bocarmé was permanently short of money. He liked to live extravagantly but earned only around 2,000 francs a year, and had therefore borrowed a substantial amount of money to fund his lavish lifestyle. It is highly probable that he married Lydie in 1840 at least in part for her money. In fact, she had little money of her own, but would inherit the family fortune in the event of her brother's death, since he was unmarried and without children. Given his poor state of health, this seemed likely to occur in the relatively near future.

However, Fougnies then dropped a bombshell. He was engaged to be married. By this time Bocarmé's finances were in a ruinous state – he owed large sums and had been forced to mortgage much of his property. The news that he and Lydie could no longer rely on receiving any money from Gustave must have come as a severe blow.

Nevertheless, the couple feigned delight at the news and invited Gustave to lunch at Château Bitremont. Unusually, their four children were sent away to eat in another part of the house on this occasion, and Lydie – a countess, let us not forget – served up the food herself. Shortly afterwards, Gustave collapsed dead in the dining room. Lydie informed the servants that her brother had suffered a stroke and vinegar was promptly poured down his throat on the grounds that it might help to revive him. It had no effect. Lydie then ordered that the servants should strip her brother naked and wash him with vinegar before removing him to a maid's room. She ordered for his clothes to be boiled and for the dining room to be scrubbed clean.

When the examining magistrate, M. Heughebaert, arrived at the scene, he ordered that the body be examined immediately. During the postmortem it was noted that Gustave's cheeks were blistered and his mouth and throat covered with burns. It was apparent from this that the cause of death could not have been a stroke, but that the deceased had somehow come to drink some sort of corrosive substance. These circumstances were considered sufficiently suspicious that Bocarmé and Lydie were placed immediately under arrest. The leading chemist in the country was Jean Servais Stas; Heughebaert sent the contents of Gustave's stomach to him in Brussels.

When he began his analysis, Stas was surprised to note the smell of vinegar coming from the sample. He was also less than satisfied when he was told that it had been administered to Gustave in an effort to revive him. There was no logic to this explanation, and Stas immediately concluded that in fact the vinegar had been used in a deliberate attempt to mask other smells that might give the real cause of death away. He therefore felt it necessary to subject the contents of Gustave's stomach to stringent testing. There were, as we know, tests to determine the presence of poisons such as arsenic in tissue; however, this required destroying the tissue itself. When the same process was used when attempting to detect non-metallic poisons, it resulted in the poison being destroyed too. He therefore had to proceed carefully.

Stas purified the stomach contents through repeated washing and filtration. He realised that the substances the stomach might contain would be soluble in either water or alcohol, but certainly not in both. Therefore, by putting a sample of the stomach contents in alcohol, he could separate the substances it contained. After purifying the contents, he mixed them with liquid ether, meaning any poison dissolved into it. Ether weighs less than water, so a layer of ether formed on top of the water. He then separated the ether from the water and let it evaporate. This left behind an oily liquid that smelled of tobacco. Nicotine in even small quantities can be a deadly poison, and Stas now began to suspect that this was how Gustave had been killed. Cautiously, he tasted a tiny amount of the substance. Not only did it taste foul, but it burned his lips, mouth and tongue. The substance was indeed nicotine. By the time he had finished extracting all of it from the stomach contents, he had enough to kill ten men.

A murder plot had been discovered despite all its imagined cleverness, and Lydie and Bocarmé were put on trial. There was considerable evidence against them: Bocarmé had an interest in both science and agriculture, and would certainly have been aware that in sufficient quantity, nicotine was a deadly poison, and also aware that as a vegetable-based poison it was supposed to be undetectable. The prosecution contended that when he had learned of Gustave's impending marriage, Bocarmé had personally extracted nicotine from tobacco leaves in order to poison him. When Gustave came for lunch, they claimed, Bocarmé and Lydie must have forced him to the floor and poured it down his throat – its strong taste would have made it impossible to put poison in his food undetected. He would quickly have lost consciousness, allowing them to administer even more of the poison. Then they used vinegar in order to mask the true cause of death. But still, the nicotine produced burns that caused suspicion to fall on the couple.

The defendants had a different story to tell. Bocarmé admitted that he had indeed distilled nicotine, but claimed to have done so out of scientific interest rather than as part of some nefarious plot. He said that vessels containing nicotine had been in the dining room, and that his wife must have accidentally used one to fill Gustave's wine glass. In short, the death had been a tragic accident. Lydie, on the other hand, claimed that Bocarmé had plotted her brother's murder, and had forced her to assist him under duress.

Perhaps unsurprisingly, the court did not accept Bocarmé's explanation of events. He was convicted and was executed by guillotine on 19 July 1851. Lydie, on the other hand, was acquitted in spite of strong circumstantial evidence against her (according to the servants, it was she who dismissed everyone from the room shortly before Gustave's death). The Stas test for detecting vegetable poisons is still used to this day – a lasting tribute to an ingenious man.

The Victorian era is generally regarded as one of poisoning's heydays, and certainly some of the most notorious cases occurred during this time. This must in large part be due to the fact that poisons of various sorts were now readily obtainable from shops – arsenic could be easily extracted from flypaper or rat poison, for example. In fact, poison became so popular as a means of murder that specific legislation such as the Arsenic Act of 1851 was brought in to try to control it. The fact that life-insurance policies were becoming more common might also have been a factor, introducing as they did a new and serious motive for murder. Bizarrely it was also not unknown for people to deliberately take arsenic in small quantities as a kind of pick-me-up or tonic (see Plate 13) – it was even regarded as having aphrodisiac properties.

The combination of these facts made the case of Florence and James Maybrick in 1889 a complicated one. Florence Maybrick (née Chandler) was born in Mobile, Alabama, to wealthy parents – her father was a partner in a bank and one-time mayor of the town. While travelling to Britain on board a ship with her mother, she made the acquaintance of a rich cotton merchant, James Maybrick. She was only nineteen, making the forty-two-year-old Maybrick twenty-three years her senior. In spite of this age gap, their relationship blossomed, and on 27 July 1881 they married in London before moving to set up home in Liverpool.

Things seemed to go well at first; the couple were well known on the Liverpool social scene and to outside observers appeared very happy. In truth, however, their home life was far from ideal, and worsened over the years. Maybrick was frequently unfaithful and had a number of mistresses, one of whom bore him five children. Perhaps prompted by the behaviour of her husband, Florence entered into several extra-marital relationships herself. One of these was with a local businessman called Alfred Brierley.

When rumours about this affair came to Maybrick's ears he flew into a violent rage. Then, apparently unconcerned at his own hypocrisy, he announced his intention of divorcing her.

However, on 27 April 1889, Maybrick was taken suddenly ill. His doctors treated him for acute dyspepsia but, in spite of their ministrations, his condition continued to worsen. He died at home on 11 May 1889. During the period of his illness, on 8 May, Florence wrote a letter to Brierley, which was intercepted by the house nanny, Alice Yapp, and sent instead to Maybrick's brother Edwin. Its contents were compromising, laying bare as they did the nature of the relationship between Florence and Brierley. As a result Edwin and the elder Maybrick brother, Michael, became convinced that Florence had murdered James so that she could take his money and be with Brierley instead.

A postmortem revealed James Maybrick's body to contain traces of arsenic, but not in sufficient quantities to have proved fatal. It was known that Maybrick had regularly used arsenic as a tonic, and indeed a city chemist confirmed that he had supplied him with it for years. A later search of the home would reveal arsenic in sufficient quantities to kill fifty people. It would therefore seem reasonable to at least admit the possibility that the arsenic found in Maybrick's system was there as a result of him self-administering.

But the Maybrick family was certain that Florence must have had a hand in the death, and after an inquest she was charged with murder and sent for trial at St George's Hall, Liverpool. One of the most significant pieces of evidence against her was the discovery that she had bought flypaper earlier that same April, and had soaked it in a bowl of water in order to extract the arsenic.

She claimed that this was with the intention of using it as a beauty treatment, which was indeed yet another common use for the deadly substance. In spite of her protestations of innocence, and in spite of the fact that the arsenic found in Maybrick's body was not a lethal dose, Florence was found guilty and sentenced to death. It is probable that the jury was swayed by the fact that Maybrick had been about to divorce Florence, which in Victorian society would ruin her. The fact that she herself had been proved to be unfaithful would also have helped cast her as the kind of 'scarlet woman' capable of committing murder.

Nevertheless, the verdict was controversial, and the case became a real cause célèbre on both sides of the Atlantic. In 1894 fresh evidence came to light – a prescription had been found inside Florence's Bible for a face wash that involved preparing arsenic. If she had arsenic readily available, it seemed strange that she would go to the trouble of extracting it from flypaper. Despite her sentence being commuted to life imprisonment, there was no possibility of an appeal. When Florence was finally released in January 1904 she returned to America, still protesting her innocence, and wrote a book entitled *Fifteen Lost Years*. She died alone and in poverty on 23 October 1941. Whatever the truth of the matter may be, this case serves to remind us that forensics cannot answer every question definitively. We will never know for sure whether Florence dosed Maybrick with arsenic or whether it was found in his body simply because he took it himself. We do know that it seems highly unlikely that the amount of arsenic detected in his body would have caused his death, and this reminds us of another important fact: that forensic evidence is useless if people wish to ignore it because it does not fit with their view of what happened.

As Maybrick's habit of taking non-lethal quantities of arsenic shows, the dosage of a particular substance is usually the crucial factor in whether it is deadly or not. And even when somebody administers a potentially lethal substance to another person with bad intent, they may not always wish to kill. It is not unknown for a criminal to incapacitate a victim with a smaller dose of a poisonous substance for the purposes of rape, robbery or kidnap. However, while a doctor is always extremely careful when prescribing drugs, and will take into account health, weight, allergies and the like, criminals have no such expert knowledge. The following case demonstrates how this can result in consequences far more tragic than the crime originally planned.

In Manchester in February 1889, a hackney carriage picked up two men, one elderly and one young. The young man ordered the driver to take them to a public house in Deansgate. When they arrived, they asked the driver if he would wait, telling him that they would not be long. Sure enough, a short while later they both emerged from the pub and directed the driver to take them to Stretford Road. While they were en route, a passer-by suddenly started calling out to the driver, trying to attract his attention. At this, he reined the horses in and stopped to find out what he wanted. The passer-by explained that he had just seen a man leap from the carriage while it was moving and disappear into a nearby alleyway. The driver climbed down and together the two of them searched for the man, but to no avail. It seemed safe to assume that this was just another case of a 'run-off' – a passenger who dashes off without paying their fare once they are near their destination. It was irritating, but not worth wasting any more time over.

Then he remembered that he still had another passenger in

the carriage; looking inside, he saw that the elderly man was still in his seat, apparently asleep. When the driver attempted to rouse him, the old man just pushed him away, without opening his eyes, insisting on being left alone. He appeared very unwell. The driver summoned the police; the constable, seeing what state the old man was in, ordered him to drive straight to the hospital, rather than to the police station. By the time they arrived, the old man was already dead.

On examining him, doctors could smell alcohol and, since there were no signs of violence on the body, concluded that he must have died of a heart attack. As a matter of routine the constable took a brief description of the young man who had leapt from the carriage – 5 feet 3 inches, clean-shaven, dressed in a brown suit and felt hat – but nothing further was done at the time.

However, the postmortem on the body discovered that the old man certainly hadn't died from a heart attack – after testing for other poisons including morphine and strychnine, they discovered that in fact he had suffered chloral hydrate poisoning. As a result the police became involved in the case once more, with Detective Constable Jerome Caminada in charge of enquiries. Born and bred in Manchester, Caminada knew the streets of the city, and its criminals, better than most. He quickly came to a number of conclusions. Firstly he deduced that, since the old man's pockets were empty, he had probably been robbed. It also seemed more than likely that the address in Stretford Road had simply been given as a blind in order to get the cabby away from the pub and to give the criminal time to get away. The victim's name, it transpired, was John Fletcher, a prominent local businessman, county councillor and justice of the peace. Why, pondered Caminada, would a man of such standing share a cab with a man who was obviously not of the same class and whom he had no reason to trust?

Caminada discovered that Fletcher had left his home that day to travel to Knutsford, where he intended to spend a few days. He had lunched in town and had arranged to meet an old friend for dinner at seven that evening. Given this, it seemed even stranger that he should have ignored these plans and instead gone to a public house with the man who was to be his murderer.

Caminada went undercover. In disguise, he searched the streets and hung out in low drinking houses, searching for any scraps of information that might help him track down the killer. He was in luck – a cab driver he met happened to remember a flash young man who had been throwing money around. He had taken him to a pub that was the haunt of boxers, gamblers, and anyone else connected to the fight game. Caminada immediately saw a connection: chloral hydrate was used by doctors as an anaesthetic, but it was also occasionally used by unscrupulous promoters as a knock-out drop in order to fix fights.

Caminada suddenly remembered the case of a criminal called Jack Parton – the circumstances of the case seemed to match his modus operandi. Parton had been a publican but had lost his licence for drugging his customers and allowing his friends to rob them while they were unconscious. Following this, he had gone into fight promotion, running crooked fights all over town. The only problem was that Jack Parton was too old to fit the description of the criminal. However, he had an eighteen-year-old son called Charlie, who did. Caminada managed to track him down and arrest him on the charge of murder, robbery and administering a stupefying drug. Charlie claimed to have an alibi, but during his enquiries Caminada discovered from a local chemist that a young man answering Charlie's description had recently been in and stolen a bottle of chloral hydrate.

Charlie was charged with Fletcher's murder and tried at Liverpool's George Hall. He claimed that it was a case of mistaken identity, but too many people had seen him – including the cab driver and the chemist. A respected witness also came forward, stating that he saw Parton pouring the contents of a phial into a glass of beer. He was found guilty and sentenced to death, though due to his youth, this was commuted to life imprisonment. There can be little doubt that Fletcher's death was a tragic accident – Charlie was certainly a far from honest young man, but it seems clear that his intention had been to drug and rob his victim, rather than to kill him. Sadly the trick he had learned from his father was a dangerous one, and in this case it went awry.

Yet another case of someone underestimating the deadliness of the substance they secretly administered occurred in England in 1954. It also serves to demonstrate the extremely unpleasant effects of yet another poison: cantharidin. Arthur Kendrick Ford was a forty-four-year-old wholesale chemist who became infatuated with two of his colleagues: twenty-seven-year-old Betty Grant and seventeen-year-old June Malins. In order to help him gain the girls' affections, he decided to use a well-known aphrodisiac called Spanish fly on them. This was a preparation made from the ground-up bodies of a particular beetle. The active ingredient is a substance called cantharidin that the beetles secrete naturally. Ford discovered that his employers kept a supply of cantharidin and he was easily able to steal some. He added a small amount to some coconut ice and then gave some to the girls, as well as eating a portion himself.

If he had expected all three of them to be overcome with pangs of lust, he was tragically disappointed. In even relatively small quantities, cantharidin is a powerful blistering agent, and is actually used in dermatology to burn off warts. Within a few hours, Ford and the girls were all taken seriously ill and rushed to hospital. Betty Grant and June Malins died shortly afterwards in agony, the drug having literally burned their insides away. Ford himself survived, though only just. When the postmortems brought to light traces of cantharidin in the bodies, Ford was interviewed and soon confessed what he had done. He was tried for manslaughter at the Old Bailey later that year, found guilty and sentenced to five years' imprisonment – a lenient sentence considering that his bizarre fantasies and stupidity cost two innocent lives.

Hycleus lugens, also known as the 'blister beetle', which secretes cantharidin.

One of the most infamous modern-day poisoners, a man who seems to have positively revelled in the practice for its own sake, is a man called Graham Frederick Young, who was born on 7 September 1947. He was fascinated with poisons from childhood, and as young as fourteen he began to poison his own family, experimenting with the different effects of varying doses. He had managed to acquire both antimony – an extremely toxic metal, which when ingested causes headaches, nausea, vomiting, dizziness and depression – and digitalis (foxglove), which is commonly used in heart medication but which can negatively affect the heart if taken in too large a quantity, as well as causing breathlessness and vomiting. He had obtained them from a local chemist by lying about his age and saying that he required them for science experiments in school.

Then, in early 1962, Young's stepmother became ill. Her condition grew progressively worse and she died suddenly in April that same year. Young's Aunt Winnie later became suspicious; she had known Young all his life and was well aware of his fascination with chemistry in general and with poisons in particular. When his father, Frederick Young, began to suffer with severe stomach cramps and vomiting, he was admitted to hospital and diagnosed with antimony poisoning. Young's chemistry teacher also found quantities of the poison in his desk at school, prompting him to call the police. Young was arrested on 23 May 1962. Under questioning he eventually admitted to the attempted murder of his father, sister and a schoolfriend. However, since his stepmother's remains had been cremated, they could not be analysed – as a result her death had to be recorded as being due to complications arising from injuries she had sustained in a car accident.

Young subsequently underwent psychiatric assessment and was found to be suffering from a psychopathic disorder. He was detained in Broadmoor Hospital under the Mental Health Act with a recommended minimum stay of fifteen years. After only nine years, however, it was considered that he had recovered sufficiently that he was no longer a danger to the public, and he was released.

In fact, although he appeared to be a model prisoner, Young busily studied medical texts during his time at the hospital, improving his knowledge of poisons. He seemingly even managed to continue to experiment on both patients and staff, one of whom (a patient called John Berridge) actually died. This went undetected at the time and it was later conjectured that, due to his expert knowledge, Young was able to extract cyanide from the leaves of laurel bushes situated in the hospital grounds.

After being discharged from Broadmoor in 1971, Young got a job at John Hadland Laboratories in Bovingdon, Hertfordshire, not far from his sister's home in Hemel Hempstead. Although his employers received references about Young's 'rehabilitation' they were not told that he was a convicted poisoner. The company manufactured thallium bromide-iodide infrared lenses, for military purposes. This might have proved convenient for Young since thallium, a heavy metal related to lead and mercury, is highly toxic. Unfortunately for him, no thallium was actually stored on site. However, once again he was able to obtain supplies of antimony and thallium by lying to a chemist, this time in London. Not long afterwards a man called Bob Egle, who was Young's foreman, took ill and died. Young was the person who was in charge of making the tea. Subsequently a number of other workers also fell ill suffering from severe nausea, some requiring hospitalisation. The outbreak was so widespread that it was at first assumed to be some kind of virus, and was nicknamed the Bovingdon Bug.

Over the next few months, Young managed to poison around seventy people, mostly using thallium obtained from the chemist. Fortunately there were no further fatalities, but quite a number of people did fall seriously ill. In all around thirty doctors were consulted during this period, but none of them recognised that poisoning was involved. This is likely due to the fact that the symptoms of thallium poisoning can quite easily be confused with those of common viruses such as influenza. Besides, the scale of the poisoning was so outrageously large that such an idea would have scarcely seemed credible. This, combined with the fact that thallium salts are odourless, colourless and almost tasteless – as well as easily soluble in water – means that in many ways it is a perfect poison. However, it is very seldom used for the purpose; indeed Young appears to have been the first poisoner to do so.

Eventually, as was inevitable, there was another fatality. Young's colleague Fred Biggs suddenly fell very seriously ill and was rushed to the London National Hospital for Nervous Diseases. He suffered in agony for several weeks before he eventually died. The company doctor noticed that Young had an unnatural interest in Biggs' death and alerted the police, who began an investigation. Young's previous conviction for poisoning immediately came to light. He was arrested in Sheerness, Kent, on 21 November 1971. When the police searched him they found thallium in his pocket, while a search of his flat turned up antimony, thallium and aconite. His diary was also recovered; it recorded the doses Young had administered to people, their effects, and whether he was going to allow an individual to live. He liked playing God.

Young was sent for trial on 19 June 1972 at St Albans Crown Court, and pleaded not guilty. Over the ten days the trial lasted, the press nicknamed him the Teacup Poisoner. However, he maintained that the diary the police had discovered was simply notes for a crime novel that he was planning to write. But there was yet more evidence against him to be presented. Naturally, in any investigation like this, establishing a cause of death through examination of the bodies of the deceased is extremely important. Such an examination had not been carried out in the case of either Bob Egle or Fred Biggs. The police had therefore obtained an order for the exhumation of Biggs' body. This was not possible in Egle's case as he had been cremated, but nevertheless police recovered the container that housed his ashes. When its contents were analysed, they were found to contain nine milligrams of thallium – a very large dose. In fact, despite all its apparent advantages, Young's choice of thallium as a poison had one flaw: although organic poisons would have been destroyed by the cremation process, a metal such as thallium survives unscathed. This became the first time in British legal history that evidence was procured from the exhumation of ashes. An autopsy on Biggs also revealed traces of thallium. On the strength of this evidence, Young was convicted and sentenced to life in prison. He died in his cell in Parkhurst in 1990, aged only forty-two.

Of course, poison is not only used by lone criminals. It has also long been used by governments as a convenient way of eliminating troublesome citizens or perceived enemies of the state, and this practice continues even today. Two classic modern cases that are often supposed to be of this type involve Georgi Markov and Alexander Litvinenko; it is alleged that they were murdered on the orders of secret services.

Georgi Markov was a well-known Bulgarian writer. The story collections *A Portrait of My Double* (1966) and *The Women of Warsaw* (1968) established him as one of the most talented young writers in Bulgaria. He also wrote a number of plays, though many of these were banned by the communist censors.

In 1969 Markov left Bulgaria to stay with his brother in Italy. It was only supposed to be a short visit, but while there he decided to remain in the West, eventually moving to London. There he found work as a broadcast journalist, working for the Bulgarian section of the BBC World Service, the American-sponsored Radio Free Europe and the German broadcaster Deutsche Welle. He had previously held a privileged position in Bulgarian society – as a talented writer he was admitted into a specially chosen intellectual group who had meetings with Bulgarian President Todor Zhivkov between 1964 and 1968. This meant he was able to reveal things that it seems certain Zhivkov would rather had remained secret; his sarcastic comments about Zhivkov probably did not go down well with the Bulgarian authorities either. In 1972 Markov's works were withdrawn from all Bulgarian libraries and bookshops and his membership of the Union of Bulgarian Writers was suspended. He was sentenced (*in absentia*) to six and a half years' imprisonment for defecting to the West.

There were a couple of failed attempts on Markov's life in 1978, one in Munich in the spring when poison was put in his drink at a dinner event, and another in the summer while he was on the island of Sardinia. The third attempt succeeded, through a method as ingenious as it was horrible. On 7 September 1978 (which happened to be the 67th birthday of Todor Zhivkov), Markov was walking across Waterloo Bridge on his way to work when he suddenly felt a sharp pain, like a tiny bite or sting, on the back of his right thigh. He spun around and saw a man picking up an umbrella from the floor – he had seemingly accidentally stabbed Markov's leg with the end of it. The man apologised hastily before running across the road and jumping into a waiting cab.

During the day, Markov noticed that a small red lump had formed where the umbrella had hurt him. He mentioned this to a few colleagues at the BBC, but otherwise thought nothing much of it. That evening, however, he developed a fever and had to be admitted to hospital. In spite of the doctors' best efforts, his condition rapidly deteriorated and three days later, on 11 September 1978, he died aged forty-nine.

The Metropolitan Police asked for a postmortem to be carried out. Markov's death had been caused by ricin poisoning. Ricin is a protein derived from castor oil and ranks as one of the deadliest known toxins – just a single gram is enough to kill around 40,000 people. A pathologist discovered a spherical metal pellet no larger than a pinhead embedded in Markov's leg; further analysis revealed it to be composed of 90 per cent platinum and 10 per cent iridium, with minute holes 0.35 mm in diameter drilled through it. Experts from the military science facility at Porton Down established that these cavities had contained ricin, and also discovered that they had been sealed with a specially designed coating. This substance melted at 37°C – the temperature of the human body. Obviously when Markov was 'accidentally' jabbed with the umbrella, he was in fact being injected with this pellet. Once inside his body, the coating gradually melted, releasing the ricin into his system. Even if the doctors had realised what had happened they could have done nothing to save him, because no known antidote exists for ricin.

Castor beans, from which ricin is derived. The poison works by inhibiting the ability of the cells to make protein. The pulp from just eight beans would cause dangerous levels of toxicity in an adult.

Although both Bulgaria and Russia were suspected of being involved in the assassination there was little that the British authorities could do. To this day, Markov's killer has still not been brought to justice, though since the fall of communism in Bulgaria there is renewed interest in the case.

The case of Alexander Litvinenko is even more recent and again involves the use of a new and extremely unpleasant poison. Litvinenko had worked for both the KGB and that organisation's successor, the FSB. In November 1998 he, along with several FSB colleagues, publicly accused his superiors of ordering the murder of a Russian oligarch called Boris Berezovsky. As a result, in 1999 he was arrested and charged with exceeding his authority. In 2000 he was released but, fearing arrest under new charges, he fled Russia with his family and was granted political asylum in Britain (having been refused it by the United States). He became a writer and journalist while also secretly working for MI5 and MI6 as a consultant.

While in London, Litvinenko wrote several controversial books. In one he accused the FSB of staging a Russian apartment bombing in which over 300 people had died, and which had been blamed on Chechen separatists at the time. He went on to claim that the FSB had been connected to other terrorist acts, saying that this had been part of a coordinated effort to bring Vladimir Putin to power.

On 1 November 2006, Litvinenko suddenly became ill and was rushed to hospital. He experienced severe diarrhoea and vomiting and became physically weaker and weaker. He lapsed in and out of consciousness. He died on 22 November, with doctors still unable to determine the exact cause of his illness (they had even suspected thallium poisoning for a time, but tests ruled this out). It was only after his death that they were able to ascertain that he had been poisoned with radioactive polonium-210. This is extremely hard to detect, since unlike most radioactive isotopes, it does not admit gamma rays, only alpha particles, which are not picked up by the majority of radiation detectors. The investigation into Litvinenko's death turned up a Russian agent called Andrey Lugovoy as a prime suspect, but when the British government requested his extradition the request was refused. Lugovoy himself denied any connection with Litvinenko's death and in turn accused the British security services. As such we will probably never know who poisoned Litvinenko, or on whose orders they acted. According to Professor Nick Priest, a Middlesex University environmental toxicologist and radiation expert, Litvinenko was probably the first person ever to die of the acute α-radiation effects of polonium-210. With his death, poisoning had entered the nuclear age.

7

DNA

DNA technology could be the greatest single advance in the search for truth, conviction of the guilty, and acquittal of the innocent since the advent of cross-examination.

Justice Joseph Harris, US judge

We have already mentioned DNA fingerprinting and its forensic applications elsewhere in the book (see Introduction). However, as it is without doubt the greatest advancement of our times in the field of forensic science, it merits further discussion. I am also going to indulge myself a little by discussing a couple of cases with which I was personally involved: one an infamous murder case from the 1980s, the other a historical mystery that I had a small hand in solving.

Colette Aram was an attractive, bright sixteen-year-old girl from the village of Keyworth, just outside Nottingham. She came from a loving home and was well thought of by her peers. She had left school to become a trainee hairdresser and by all accounts loved what she did and was on course to make a success of it. This was not to be. Shortly after 8 P.M. on 30 October 1983 (I will always remember the date because it was my birthday), she left her home on Normanton Lane in Keyworth to walk to her boyfriend Russell Godfrey's house, just over a mile away. Normally he would have picked her up in his car but, as often seems to be the case when such awful events occur, fate took a hand – on the night in question his car was off the road. Colette was last seen a little after 8 P.M. talking to friends at a junction between Nicker Hill and Platt Lane. She was walking in the direction of Willow Brook. A witness later reported having heard someone screaming and then a car driving off at high speed not long afterwards. He looked out onto the street but saw nothing and, since he was used to hearing children shouting and screaming nearby, did not think it unusual or report it.

When Colette hadn't arrived at Godfrey's house by 10.30 P.M. that evening, people started to become concerned. Finally, the police were called. It was a freezing cold October night and a hard frost was already beginning to set in. After several fruitless hours the search was called off until the next day. In fact, continuing it proved horribly unnecessary; the following morning a man driving to work along Thurlby Lane, less than two miles from where Colette lived, spotted something strange-looking in a field beside the road. Concerned, he turned his car around and went to investigate. What he had found was Colette's naked body. She had been beaten, sexually assaulted and then strangled. She had been dead for several hours.

The murder investigation began at once, headed up by Detective Superintendent Bob Davy. An incident room was established on the Keyworth and Normanton playing fields, close to Nicker Hill where Colette had last been seen.

Hundreds of police officers, including myself, were drafted in for a vast and wide-ranging inquiry. As a result, as is frequently the case with major inquiries, other previously unreported crimes – some extremely serious – came to light, and several people were arrested for offences unconnected to Colette's murder.

It transpired that a red Ford Fiesta had been stolen earlier on the day of the murder from the stables at Holme Pierrepont, another small village in the area. This car was then discovered abandoned in Keyworth, while its keys were subsequently found hidden in a bush. When Colette's body was found, her feet were free from mud and there were car tyre marks leading out of the field. This suggested that any assault on her had probably taken place inside a vehicle, so the discovery of the car seemed highly significant to the murder squad. A thorough forensic examination of its interior turned up traces of both blood and semen.

Two girls now came forward, each to report that she had been followed by a man in a red Ford Fiesta on the night Colette disappeared. One had felt so threatened that she had actually run into the house of a friend. The other, who was fourteen years old, had been approached by the man, but the fact that she had a large dog with her was obviously enough to make him think twice, and he drove away. Both girls gave a similar description of the suspect: a white male about 5 feet 10 inches tall, with dark wavy hair. This was definite progress for the inquiry. However, even more promising was information from the landlady of the Generous Briton public house in the village of Costock. A man had visited the pub at approximately 9 P.M. on the night of the murder. He had ordered an orange squash and she had noticed blood on his hands. When she pointed this out he went into the toilets to clean them. This was highly fortunate; following a search of the toilet, a bloodstained paper towel was recovered and kept as possible evidence. It was later to prove extremely significant.

On 7 June 1984, the case became the first ever to be featured on the now long-running BBC crime show *Crimewatch*, a fact that delighted the senior investigating officers. The programme received over 400 calls offering information connected to the case, as a result of which the murder squad were able to eliminate over 1,500 suspects from their lists. Still, in spite of all of this useful information, they seemed no nearer to actually capturing the killer.

On 17 November 1983, the incident room received a letter bragging about the murder. It revealed things that only the killer could have known, and taunted the police, saying that they were never going to catch him. The letter was thoroughly examined and a fingerprint was discovered on it. Unfortunately, any hopes the police might have had of this finally leading them to the killer were soon quashed; there was no match for it on their database.

And so, despite a lengthy inquiry involving hundreds of police officers, the culprit remained at large. There is always a strong feeling of gloom and failure within a team when, in spite of everyone's best efforts, a culprit eludes them. It was no different with the Colette Aram murder squad. There was a sense of anger and frustration at not having got our man. Gradually the investigation was run down – it remained on file with the odd piece of information being followed up on from time to time, but to all intents and purposes the case seemed to be dead in the water. The years rolled by. The case was featured yet again on *Crimewatch* for their twentieth anniversary show in 2004, but despite this once again prompting a flurry of phone calls, no useful new leads came out of it. The memory of Colette and the feeling of failure attached to it lingered with me personally for the next twenty-five years. I've often thought how much worse it must have been for the senior investigating officers. So that was it: just another unsolved murder.

But science had not been idle all those years, and twenty-five years after Colette's death, forensic technology had exceeded even the most optimistic police officer's wildest dreams. DNA had burst upon the scene when Alec Jeffreys from the University of Leicester, along with Peter Gill and Dave Werrett of the Forensic Science Service, developed DNA profiling or genetic fingerprinting. We have already seen how this work came into its own during the infamous Black Pad murders. I would later work with Peter Gill in connection with a major historic mystery (on which more later) and he explained his own part in the development of the technology as follows: 'I was responsible for developing all of the DNA extraction techniques and demonstrating that it was possible after all to obtain DNA profiles from old stains. The biggest achievement was developing the preferential extraction method to separate sperm from vaginal cells – without this method it would have been difficult to use DNA in rape cases.' These techniques were, of course, highly applicable in the case of Colette Aram.

In fact, in 1997 Colette's murder was reviewed and an attempt was made to establish a DNA profile of the killer. Dr Tim Clayton managed to reconstruct Colette's DNA from samples kept after her postmortem. However, his attempts to provide a DNA profile of the killer from the samples of semen found in the car and on Colette's clothes were only partially successful. Unfortunately, to get a full profile, he needed twenty markers, ten from the father and ten from the mother – he was only able to establish three. (DNA markers are sections of DNA that are unique to an individual so can be used to identify them. They are found at particular locations on the DNA molecule.) Clayton certainly did not have enough to recognise the killer, though it did help to eliminate a number of other people of interest from the inquiry. Still, the 1997 review was finally wound down, with the case being not much further forward than it had been in 1983.

In 2004 Detective Superintendent Kevin Flint, who had only been a detective constable at the time of Colette's murder, became head of the Nottinghamshire homicide unit. Like the rest of the team who had worked on the case in 1983, he was disappointed that the killer had still not been brought to justice, and was determined to put this right – perhaps unsurprising considering his reputation for dogged persistence.

Once again the evidence was reviewed, and once again Tim Clayton was asked to examine the evidence. This time, however, advances in DNA techniques allowed Clayton to work on the paper towel found in the toilets of the Generous Briton pub. Developments in 'low copy' DNA, in particular, now enabled scientists to build a full DNA profile from a very small number of DNA markers. Clayton thought it was a long shot, since they had no idea who the man in the pub had been, never mind whether he was involved in Colette's murder. But when he examined the towel he discovered that it contained DNA from two separate people, one male and one female. He immediately compared the female sample with the known DNA profile of Colette Aram. It matched. This was enormously important: it almost certainly meant that the man who had been in the pub that night was Colette's killer. And just as significantly, Clayton was next able to extract a full sample of this man's DNA. This matched the three markers he had managed to obtain back in 1997. The team now had a full DNA profile of their killer.

That profile was then searched for against the police database. Hopes were high, since surely a man capable of murder in 1983 might well have committed further offences in the years since? But the results were another disappointment – there were no matches. Still not admitting defeat, Clayton suggested the idea of doing a familial DNA search. Such a search highlights anyone whose DNA bears similarities to the sample in question, and who therefore might be related in some way to the person from whom the sample comes. The results were daunting, to say the least: initially the system turned up thousands of possibilities. With some work, however, this was eventually cut down to approximately 300 people of interest. This was obviously an improvement, but even this smaller number would take many months to examine.

A further eighteen months went by, and after a great deal of footwork all 300 or so suspects were eliminated from the inquiry. It seemed that in spite of all the apparent progress the team had made, the inquiry would once again have to be wound down as it had been in 1983 and 1997. This was a bitter pill to have to swallow. In a last-ditch effort, Clayton suggested that they run through the DNA database one more time to see if any of the new profiles that had been added over the last eighteen months while they were concentrating on their enquiries were a match. It felt a bit like clutching at straws, but they had nothing to lose.

To Clayton's amazement, when he ran a familial DNA search again, a match suddenly appeared at the top of the list. It was a man called Hutchinson, who had been arrested on a motoring offence in 2008 and had given a sample of DNA. Although it was similar to the sample from the killer, it was not identical; besides which he was only twenty years old, so it was quite impossible that he could have committed the offence. Still, Clayton was certain that he had found a close relative of the man they were hunting.

Kevin Flint and his team began to investigate the Hutchinson family. The man whose DNA they had matched was the son of a man called Paul Hutchinson. He was in his fifties at the time of the renewed investigation and was one of four brothers (one of whom had recently died), who had all lived near to the stables at Holme Pierrepont. Paul had moved to Keyworth in early 1983. The decision was made to arrest all three of the surviving brothers, and on 7 April 2009 they were taken into custody.

Although his two surviving brothers co-operated fully with the police, Paul decided to remain silent when questioned. DNA samples and fingerprints were taken from each of the men and, when tested, Paul's profile was a positive match with that of the killer. His fingerprint likewise matched that found on the goading letter sent to the police years before. Hutchinson at first denied everything and tried to blame his dead brother for the murder. However, the team managed to obtain a DNA sample for this brother from the hospital where he had died, and were thereby able to definitively disprove Hutchinson's story. Faced with the mounting evidence against him, he finally admitted his guilt. He was sentenced to life imprisonment with a minimum of twenty-five years to be served. When sentencing him Mr Justice Flaux described the attack on Colette as 'truly horrendous', saying that 'the terror and degradation that this poor girl must have suffered at the hands of a stranger in her last few moments are unimaginable'.

The Association of Chief Police Officers later said that the case demonstrated the importance of the controversial national DNA database, which has been criticised by civil liberty campaigners. A spokesperson for the association commented that 'DNA continues to help resolve a substantial number of crimes by either detecting those responsible or eliminating people from police enquiries. This case is yet another example of the value of DNA evidence.' Whatever your feelings on the existence of a database like this, it's hard to argue against its effectiveness in criminal detection. The fact that modern advances mean that even very old genetic material can still provide viable DNA samples is also an incredible step forward in forensics, and meant that an evil man was caught twenty-five years after his crime. There are doubtless other people out there now who, like Hutchinson, thought they had escaped punishment for their crimes. Now, thanks to this technology, they can no longer feel so secure (see Plate 14).

The story has one more bizarre turn to take. On 11 October 2011, Hutchinson was discovered unconscious in his cell at Nottingham Prison. He died shortly afterwards in the ambulance on his way to hospital. The cause of death has never really been established; some say it was a heart attack, others suicide. Whatever the case may be, there can have been few who mourned his passing.

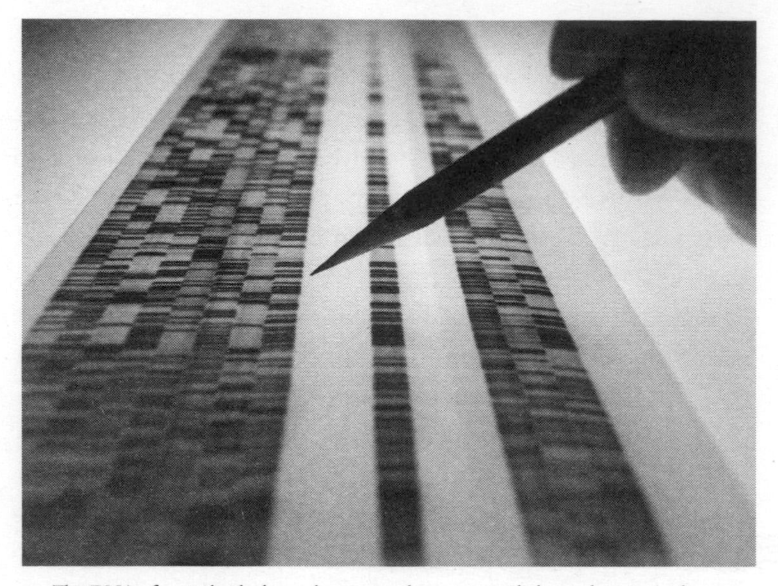

The DNA of an individual may be separated out using gel electrophoresis, as shown here, and analysed for DNA markers. Twenty of these markers can positively establish the identity of an individual.

One slight problem with the ceaseless march of science is that it can be very difficult for the courts and judiciary to keep up with the speed of forensic advances. Even the most learned of judges and the most attentive of juries can be left confused and overwhelmed by the science involved, struggling to interpret its complexities.

DNA fingerprinting was first offered in evidence in the United States in 1988 during the trial of the rapist Tommy Lee Andrews. Andrews lived in Orlando, Florida, and worked at a pharmaceutical warehouse. His vicious attacks began in May 1986, when he raped twenty-seven-year-old Nancy Hodge, a computer operator at Walt Disney World Florida. He attacked her from behind while she was in her bathroom, threatening her with a knife and covering her face during the assault to prevent her from identifying him later. He violated her three times before leaving the apartment, taking her bag with her. This was a pattern that he was to stick to over the course of his other attacks: he always covered his victim's face and he always took a personal item from them. He continued to assault and rape women with alarming frequency, so that by December 1986 he had struck an appalling twenty-three times. He was careful and covered his tracks well, meaning that for a long time there seemed little hope of catching him. That is until February 1987, when he attacked Karen Munroe in her own home in Orlando. Despite the brutality of the assault, she was fearful of waking her children and putting them in danger, so she did not scream. On this occasion Andrews left two fingerprints behind. The police stepped up their surveillance operations and in March 1987 they got lucky. A woman in the local area reported seeing a prowler and the police, reacting quickly, arrived just in time to see a blue 1979 Ford racing away from the scene. They pursued the car for several miles before the driver eventually lost control and crashed. That driver was, of course, Tommy Lee Andrews. It was soon discovered that his fingerprints matched those discovered at the scene of the latest crime, and he was immediately charged with rape. However, although the police were pleased to have him in custody, they wanted to be able to link him to the numerous assaults they were certain he had committed, not just one. Of all of his other victims, only one recognised him, and even she could not say that it was him with absolute certainty, since she had only glimpsed her attacker. Andrews' blood type matched the blood type of the man from whom the semen samples taken from the victims had come, but then the same was true of 30 per cent of the male population. A good defence lawyer would tear such evidence to shreds. The police needed more in order to successfully charge Andrews with serial rape.

The inquiry team had heard about the British success with DNA fingerprinting, especially in the case of the Black Pad murders (see Introduction). With this in mind they contacted Lifecodes, a DNA testing laboratory based in New York. Samples of blood and semen were forwarded there and one of their scientists, Dr Alan Giusti, set to work. The results were returned a few months later. The genetic fingerprint of Andrews' blood sample, taken while in custody, and the genetic fingerprint of the semen samples taken from the victims were identical. They came from the same man.

At his trial for raping Nancy Hodge, Andrews pleaded not guilty, stating that he was at home at the time of the attack. His story was backed up by his girlfriend and his sister. The DNA evidence – which had been allowed by the court – was then presented. The prosecution pointed out that the chances of a false match – of the DNA from the scene not coming from Andrews – were ten billion to one. One might think that, given this, securing a conviction was all but certain. However, this was a new kind of evidence, one that nobody was used to dealing with. The defence very shrewdly countered the prosecution by asking them to prove that what they were saying was true. They were caught completely flatfooted, without any data on hand to back their assertions. As a result the jury was split and a mistrial declared.

Two weeks later, Andrews stood trial for another rape, that of Karen Munroe. This time, though, the prosecution lawyers were ready – on top of DNA evidence, they also had fingerprints from the scene. Andrews was found guilty and sentenced to twenty-two years' imprisonment. The retrial for the Nancy Hodge case followed soon after. Andrews stuck to his story, and his sister and girlfriend continued to support him. In the end it all came down, once again, to the DNA evidence. With great care, keeping the explanation as simple as possible, the prosecution explained how the process of DNA profiling worked, why the samples taken from Andrews and from the rape victims matched, and why that meant that Andrews had to be the culprit. The defence once again tried to discredit this evidence with pseudo-science, but this time they failed. Andrews was found guilty again, this time of serial rape. After his sentence was extended, his total jail time stood at 115 years. DNA had at last proved itself as an admissible and valuable form of evidence in the United States, as it had already done in Britain. It is now widely used, and while it may occasionally be challenged in court on the grounds of contamination or technical ineptitude, the science itself is beyond question.

Although it may seem remarkable that it was possible for the Hutchinson murder to be concluded through DNA evidence twenty-five years after the crime itself was committed, in fact the same forensic techniques can be used to establish identity even longer after the deaths of those concerned. This is the case, for example, with Tsar Nicholas II and his family, who were killed at Yekaterinburg in 1918 (see Plate 15).

By March 1917, Nicholas Romanov was no longer tsar, having abdicated in favour of his brother Grand Duke Michael. He and his immediate family were placed under house arrest at the Alexander Palace in Tsarskoe Selo while the provisional government, under the control of Alexander Kerensky, decided what should be done with them. Many of their other relations escaped Russia and fled into Europe, where their descendants live to this day. In August 1917, the Romanov family were moved to Tobolsk, the traditional capital of Siberia, supposedly to protect them from revolutionary violence. They were well looked after and lived in some comfort, occupying the former governor-general's house. But when the Bolsheviks came to power in October 1917, the situation for the family became more serious, not to mention stricter. They were forced to dismiss most of their servants and were placed on soldiers' rations, giving up the last of their luxuries such as chocolate and butter. Then, during the summer of 1918, the Bolsheviks moved the family to Yekaterinburg, where they were imprisoned in Ipatiev House (also known as the House of Special Purpose), located at 49 Voznesensky Prospekt.

The civil war was still raging, and with the Czechoslovak Legion closing in on the city, the Bolsheviks decided it was necessary for them to murder the Romanov family in order to prevent the White Russians (those with tsarist sympathies) rallying around them. Yakov Sverdlov, a Bolshevik party leader and chairman of the All-Russian Central Executive Committee, signed the telegram ordering the execution of the family, though it is certain that Lenin himself would have had the final word on such an important decision.

Trotsky later referred to their deaths in his diary: 'My next visit to Moscow took place after the fall of Yekaterinburg. Talking to Sverdlov I asked in passing, "Oh yes, and where is the Tsar?" "It's all over," he answered. "He has been shot." "And where is his family?" "And the the family with him." "All of them?" I asked, apparently with a touch of surprise. "All of them," replied Yakov Sverdlov. "What about it?" He was waiting to see my reaction. I made no reply. "And who made the decision?" I asked. "We decided it here. Ilyich [Lenin] believed that we shouldn't leave the Whites a live banner to rally around, especially under the present difficult circumstances."'

From what we know, at around midnight Yakov Yurovsky, commander of the house guard, woke the Romanovs' doctor Eugene Botkin and ordered him to instruct the family to dress and to assemble in a small basement room. He was to tell them that they were to be moved from the house for their own protection due to trouble in the town. When all the family were gathered, Yurovsky entered the room along with an execution squad and read out the order given him by the Ural Executive Committee – the order to kill the family. The squad then opened fire. Nicholas himself was shot in the head and fell to the ground, though several of the children were not killed in the opening fusillade since they had a considerable quantity of diamonds sewn into their clothes, which protected them to some extent. They had to be finished off with bayonets. In twenty minutes, it was all over.

For years afterwards the location of the bodies of the tsar and his family was a secret. Many thought that the bodies must have been thrown down a mineshaft or burned – perhaps both. Books and articles were published claiming that the family were alive and well and living in Siberia. Over the years the lack of bodies meant a great many people were able to come forward purporting to be Romanovs, mostly to try to claim the royal fortune, which was said to be hidden in banks across Europe. The most famous of these impostors was Anna Anderson, who claimed to be Anastasia, the youngest daughter in the family. Meanwhile plenty of other people who accepted that the Romanovs were dead were trying to find their last resting place. As you might imagine, under a communist government this was a difficult, not to mention dangerous, endeavour.

A geologist named Alexander Avdonin was particularly committed to the search. He lived in Yekaterinburg, and was a keen amateur archaeologist with an interest in local history. The murder of the Romanov family was a story that particularly fascinated him. Avdonin researched the incident for years, slowly collecting evidence on what might have happened to the bodies of the royal family after they had been shot. In 1976, as a result of his sustained interest, he met writer and filmmaker Geli Ryabov, who had been given some information by the son of one of the men involved in the shooting and was certain that he knew where the tsar and his family were located. According to Ryabov's source, nine of the eleven bodies were buried near grade crossing 184 on Koptyaki Road. Supposedly, after the bodies had been dumped into the grave, acid had been poured over them to help destroy them. After that, railway sleepers were placed over the hole before earth was thrown on top.

Armed with these clues, Avdonin and Ryabov began to search for the spot in the spring of 1979. Their luck was in, and it was not long before they struck the rotted wood of the sleepers. Shortly afterwards they discovered broken fragments of what they took to be the jars that had contained the acid. Spurred on by this, they kept digging and finally unearthed several skulls. They were now all but certain that they had found the long-lost royal family of Russia. Suddenly the enormity of this discovery dawned on them, and they began to think about the potential repercussions. Worried about what might happen, they reburied the remains along with a number of icons. They were to keep their secret for ten years. Then, in 1989, Ryabov released the story to the media.

Eventually the bodies were exhumed again, this time officially. Given the location in which they were discovered, the fact that one of the skulls had golden bridgework (which Nicholas Romanov was known to have had) and the fact that when that same skull was superimposed against a photo of Nicholas it matched perfectly, it seemed highly probable that these truly were the bodies of the Romanovs. However, as compelling as the circumstances might be, there was still no absolute proof, and so doubts lingered.

It is at this point that my involvement in the story begins. I came across the story of the remains while I was working on the BBC show *Tomorrow's World* in 1992. Fascinated, I contacted the Russian Forensic Science Service in Moscow to find out more, and was put through to one of their leading DNA specialists, Dr Pavel Ivanov. He outlined what they had done so far, but explained that they had insufficient funds to bring the remains to England for DNA analysis. I offered to pay, which he was delighted about. The next step was to contact the then Home Secretary (and fortunately my own

The cathedral of St Peter and St Paul in St Petersburg, where the remains of the Romanov family were finally laid to rest.

constituency MP), Ken Clarke, who authorised the DNA work to be carried out at Aldermaston, the UK Home Office's forensic science service. It was to be done by a British scientist, Dr Peter Gill, who was extremely excited at the prospect. Pavel Ivanov then flew into the UK carrying nine right arms in an old British Airways bag. The bag promptly went into the back of my Volvo and we drove to Peter Gill's house. I couldn't help but wonder how the police might react if they stopped me and found that I had nine skeletal arms stashed away. Still, there can't be many people who've had an entire royal family in the boot of their car.

The genetic analysis of the bones was carried out over a number of weeks, testing against samples of relations to the Romanovs, such as the Duke of Edinburgh, whose maternal grandmother Princess Victoria of Hesse and by Rhine was the sister of Tsarina Alexandra. At the end of this period Peter Gill had established that the remains were indeed those of the family. It was momentous news, and obviously received a great deal of media attention. Given that there was no longer any doubt about who they were, the remains of the Romanov family were also finally able to be given a proper burial. They were interred in the vaults of the cathedral of St Peter and St Paul in St Petersburg on 17 July 1998. There they now lie, along with so many great tsars of Russia. I subsequently received a letter from the Russian Forensic Science Service thanking me for my help in the project, and also received the thanks of the surviving members of the Romanov family – I am very proud of both.

DNA fingerprinting also ended the question of Anna Anderson's controversial claims to have been the Grand Duchess

Anastasia. Even as early as 1927, a private investigation funded by Tsarina Alexandra's brother Ernest Louis, Grand Duke of Hesse, had identified Anderson as Franziska Schanzkowska, a Polish factory worker with a history of mental illness. But without stronger evidence there was no way to definitively debunk her claims. In fact, it would not be until after her death in 1984 that the question could finally be settled. It turned out that part of Anderson's intestine, which had been removed during an operation in 1979, had been stored at a hospital in Charlottesville, Virginia, where she had lived out her final years. Analysis of the DNA from this not only proved that this woman was not related to the Romanovs, but was also able to match her with a sample given by Karl Maucher, a great-nephew of Franziska Schanzkowska. It seemed that the original investigation had been right all along. There might be a certain romance to the notion that Anastasia had survived all those years that made some people willing to believe it, but in the end DNA analysis revealed the truth.

But if it seems shocking that science can reach so far back into the past, perhaps the most astounding piece of recent DNA analysis reaches back even further – hundreds of years, in fact. King Richard III was born on 2 October 1452. He reigned as king for two years, from 1483 until his death at the Battle of Bosworth Field in 1485. He has the distinction of being the last king of the House of York, the last of the Plantagenet dynasty. His reign is now somewhat infamous and he is often vilified, as in Shakespeare's *Richard III*.

When Richard's brother Edward IV died in 1483, Richard became Lord Protector on behalf of Edward's son, the twelve-year-old King Edward V, along with his brother Richard. Richard had

A portrait of Richard III by an unknown artist, thought to have been painted before 1626. He is suspected to have suffered from scoliosis, which would have made one shoulder appear higher than the other, as does indeed seem to be the case in this painting.

the two boys housed and locked away in the Tower of London for, Richard claimed, their own protection. Edward's coronation date was set for 22 June 1483. However, before the young king could be crowned, his father's marriage was declared invalid because of a prior union, thus making his children illegitimate and ineligible for the throne. This was extremely convenient for Richard, who was therefore able to ascend to the throne himself.

During his reign, Richard had to confront two major rebellions. The first, in October 1483, was led by the Duke of Buckingham and supporters of Edward IV, who believed Edward's sons were the true heirs to the throne. Richard crushed the revolt and Buckingham was executed. Then, in August 1485, Henry Tudor and his uncle Jasper Tudor moved against Richard. The two armies finally met at Bosworth Field. At first it seemed that Richard would defeat Henry, since his army was considerably larger. However, Richard was killed while leading a cavalry charge towards Henry in an attempt to cut him down and finish the battle. He was the last King of England to be killed in battle.

It is recorded that Richard III's remains were buried by friars in a nearby church. However, for many years a legend persisted that, not long after his death, his body was taken from its grave in Greyfriars cemetery, thrown into a river by an angry crowd and was therefore lost for ever. No evidence has ever been found of this.

The fact that the search for the site of Greyfriars church got under way was largely due to a member of the Richard III society, Philippa Langley, with Dr John Ashdown-Hill and Annette Carson, both of whom had written books about Richard. Their first task was to try to raise the money they needed in order to begin any serious exploration. With the help of the members of the Richard III Society, they not only reached but considerably exceeded the sum of £10,000 which they needed. Then Richard Buckley, the lead archaeologist at the University of Leicester – it is really rather surprising how important the University of Leicester has been to the history of forensic science – became involved in the project. Although he considered that the chances of actually finding the body of the king were slim to none, he was willing to have a try; after all, they might find something else of interest on the site, even if they could not locate the royal remains.

An eighteenth-century map of Leicester showed that the site of Greyfriars church was now under the office of Leicestershire Social Services. On 25 August 2012, the team began to excavate in the car park of the building. By 12 September that year they were able to make an extremely exciting discovery public: they had discovered the skeleton of an adult male. Although it was of course too soon to make a real identification, there were things that made the team dare to hope that they had found the lost body of the king.

For example, when the team used a CT scan on the remains to produce a 3D record of every bone in the body, it was established that the skeleton showed signs of scoliosis – a slight curvature of the spine. While this would not have made the man a hunchback of the sort Richard is sometimes represented to have been, it would likely have made one shoulder visibly higher than the other, and might be the kind of feature that detractors would have picked up on and exaggerated for their own propaganda.

The skeleton also exhibited various war wounds – no fewer than ten, in fact. It seemed that this man had lost his helmet at some point in battle, since most of these were to the skull.

There was a stab wound, almost certainly made by a rondel dagger, which was a popular weapon at the time; there was a slicing injury that must have been caused by a flat-bladed weapon, and finally there was a massive cleave to the back of the skull that would almost certainly have exposed the brain. It was this last injury that must have been the fatal one. At any rate, it was clear that this man had died in battle.

As much as everyone wanted to believe that the skeleton might truly be Richard III, all this was only compelling circumstantial evidence. But DNA analysis might be able to provide more definite proof. And Dr John Ashdown-Hill had managed to do something really quite extraordinary which would assist in this effort. Through in-depth genealogical research he had succeeded in tracing a distant descendant of Anne of York, Richard III's older sister: a British woman named Joy Ibsen who had emigrated to Canada shortly after the Second World War. She was a sixteenth-generation great-niece of Richard, along a direct maternal line. This last point was significant because it allowed her to be used for the purposes of mitochondrial DNA analysis. Mitochondrial DNA represents a tiny amount of the total genetic material that any one of us carries, and is distinct from that carried in the main chromosomes. Every person inherits their mitochondrial DNA from their mother, without the usual recombination of genes from mother and father, meaning that along a direct maternal line, every person will have the same mitochondrial DNA.

Sadly Joy died in 2008, so it was her son Michael Ibsen who gave a mouth-swab for the purposes of comparison in 2012.

His mitochondrial DNA was found to belong to DNA haplogroup J; if the remains were those of Richard III, their mitochondrial DNA would be from this haplogroup as well. The work on the remains was carried out by geneticist Dr Turi King, who was able to confirm that this was indeed the case. Of course this wasn't an absolute identification, as a large number of other people might belong to haplogroup J, but given the location of the remains, their age, their physical appearance and the injuries to them, this extra piece of evidence seemed to confirm beyond reasonable doubt that the body was that of King Richard III, and on 4 February 2013, the University of Leicester announced this to the world. In an ending comparable to that of the discovery of the Romanov family, it has since been announced that Richard will be properly interred in Leicester Cathedral in early 2014.

Finally it is worth looking at a case that demonstrates better than any other the capacity for DNA evidence to exonerate the innocent. It should surely give pause to advocates of the death penalty, since it demonstrates beyond doubt that miscarriages of justice occur in such cases.

In 1984 a nine-year-old girl called Dawn Hamilton was raped and murdered in Rosedale, Maryland. A former marine named Kirk Noble Bloodsworth was arrested and charged with the crime, the main evidence against him being several eyewitnesses who placed him with or near Dawn around the time of her murder. The prosecution also maintained that footprints found on the victim's body matched a pair of shoes found in Bloodsworth's home. He steadfastly protested his innocence but in 1985 he was found guilty and sentenced to death. Luckily, in 1986 it was discovered that the prosecution had illegally withheld evidence from the defence and the Maryland court of appeal overturned the original conviction. When Bloodsworth was tried again he was found guilty once more, but this time was sentenced to two terms of life imprisonment rather than death. This very possibly saved his life, as had he remained on death row, his sentence might well have been carried out before he had a chance to prove his innocence.

However, in 1992 Bloodsworth happened to read about the Black Pad murders and how DNA evidence had both eliminated an innocent man from the investigation and led to the capture of the real perpetrator. He saw the potential of this new technology for his own case immediately. Later that year, he succeeded in obtaining a court order for the testing to occur. Initially it looked as though this would be impossible, as the DNA evidence that existed from the case – Dawn Hamilton's underwear, which contained traces of semen from her attacker – could not be found. Eventually, however, it was located in an evidence bag in the judge's office. The necessary analysis was carried out in Richmond, California, at Forensic Science Associates. The results completely exonerated Bloodsworth and on 28 June 1993 he became the first death-row prisoner to be released on the basis of DNA evidence – a wonderful forensic landmark. In 1995 the governor of Maryland, William Donald Schaefer, granted him a full pardon.

Ten years later, in the kind of extraordinary twist that we have by now come to expect in the world of forensics, additional DNA evidence that had been added to state and federal databases led to the identification of the real killer in 2003. He was a man called Kimberley Shay Ruffner, and had in fact been incarcerated on unrelated attempted rape and assault charges just one month after Bloodsworth had been sentenced. Astonishingly, he was housed in a cell on the floor below Bloodsworth, and the two knew each other well; Ruffner worked in the prison library and used to bring Bloodsworth his books. Ruffner was charged with Dawn Hamilton's murder and in 2004 pleaded guilty to the crime for which Bloodsworth had been wrongfully convicted. Bloodsworth was paid $300,000 in compensation and is now unsurprisingly a vocal supporter of measures such as the Innocence Protection Act, which seeks to minimise the risk of innocent people being executed.

The power of DNA fingerprinting is quite simply astonishing – particularly when set against the early methods of identification described in this book. It is impossible to overstate the importance of DNA technology in the field of criminal investigation. To be able to definitively identify a person and link them to a crime scene is a power so striking that it seems almost fictional – the deductions of Sherlock Holmes and Hercule Poirot pale in comparison. To observe these techniques being used to reach back into history and solve mysteries that have remained unanswered for centuries only serves to underline the incredible potential of this technology.

In examining the history of forensic science it is inevitable that we are forced to confront the darker side of human nature. We look at these brutal crimes and we say to ourselves in disbelief, 'How could anyone do that?' I cannot disagree – it is impossible not to look at such acts and see evil. But I hope that through explaining the complex and painstaking methods by which these crimes are solved, this book has also demonstrated that forensic science instantiates much of what is best in humanity: ingenuity, determination and above all a belief in justice.

Picture Acknowledgements

Plate 3 © Everett Collection Historical / Alamy
Plate 6 © BSIP SA / Alamy
Plate 7 © Martin Phelps / Alamy
Plate 8 © Scenics and Science / Alamy
Plate 9 © Louise Murray / Alamy
Plate 10 © Mark Bourdillon / Alamy
Plate 11 © B Christopher / Alamy
Plate 12 © Mary Evans Picture Library / Alamy
Plate 13 © PARIS PIERCE / Alamy
Plate 14 © Mark Harvey / Alamy
Plate 15 © Archive Pics / Alamy
Figure 1 © The Print Collector / Alamy
Figure 2 © fStop / Alamy
Figure 4 © Art Directors & TRIP / Alamy
Figures 5, 7, 9, 12, 15, 19 © Alamy
Figure 10 © Old Paper Studios / Alamy

All other images courtesy of Wikimedia Commons.

Index